STARTING SOMETHING BIG

STARTING SOMETHING BIG

The Commercial Emergence of
GE Aircraft Engines

R. V. GARVIN

American Institute of Aeronautics and Astronautics, Inc.
1801 Alexander Bell Drive
Reston, VA 20191

Publishers since 1930

American Institute of Aeronautics and Astronautics, Inc., Reston, Virginia

1 2 3 4 5

Library of Congress Cataloging-in-Publication Data

Garvin, R. V. (Robert V.), 1927–
   Starting something big : the commercial emergence of GE aircraft engines / R. V. Garvin.
     p. cm.
     Includes index.
     ISBN 1-56347-289-9 (alk. paper)
     1. GE Aircraft Engines (Firm)—History. 2. Aircraft engine industry—United States—History. 3. Airplanes—Turbojet engines—History.
I. Title
HD9711.2.U64G414   1998   629. 134'353'0973—dc21   98-51883

Cover design by Sara Bluestone

Copyright © 1998 by the American Institute of Aeronautics and Astronautics, Inc. All rights reserved. Printed in the United States of America. No part of this publication may be reproduced, distributed, or transmitted, in any form or by any means, or stored in a data base or retrieval system, without the prior written permission of the publisher.

Table of Contents

Preface	vii
Introduction	ix
1. How GE Got into the Jet Engine Business	1
The First Generation: 1903 to 1945	1
Jet Power and the Cold War	7
Military Technology Flows into Commercial Jets	14
CJ805 Service and Product Support	19
2. Business Jets	23
3. The Years of the Cold War	31
4. Back in the Commercial Engine Business	39
5. Airbus Industrie and the Birth of Revenue-Sharing Strategy	53
6. Engines for the Boeing 747	61
7. The Big Twins Reshape Competition	65
8. The Military Market Share	77
9. Competition as a Factor in the Military Market	97
10. License Agreements and Offset Programs	107
11. The Foreign Corrupt Practices Act	119
12. The Gap in Everyone's Commercial Product Line	125
13. But Will It Sell?	133

14.	The Japan Factor	143
	Military Aircraft Production in Japan Sets the Stage	149
	Commercial Engine Experiments	151
	The Real Product	153
15.	Advanced Technology Programs	161
	Quiet Clean Short-Haul Experimental Engine	162
	Energy Efficient Engine	165
	The Unducted Fan	167
16.	Developing the Chinese Market	173
	An Old China Hand	177
	Learning the Ropes	179
	The Peking Hotel	181
	Missteps by the Chinese Industry	184
	Shenyang	189
	Making Parts in China	194
	The Liming Visit to America	200
	The Missing Engine	203
	Setting Up in Beijing	206
	The New Team	221
	The Zong Shan Affair	224
	Tienanmen	232
	Winding It Up	236
17.	Developing the Russian Market	239
18.	Product Planning and Partnership—Mixing Oil and Water	257
19.	The New German Engine	269
20.	Product Diversification	279
21.	Service as a Separate Business	294
22.	The Death of Douglas	301
23.	The Cyclical Airline Industry	309
24.	Lessons for the Investor	317
25.	Are There Any Lessons for the Russians?	329
26.	The Bottom Line	337

Preface

Advances in aviation and propulsion technology over the past 30 years have revolutionized airline transportation. Today's global airline system was made possible by the creativity and technical expertise of many key individuals at GE Aircraft Engines (GE), Pratt & Whitney, and Rolls Royce. Through the creative efforts of many engineers and the entrepreneurship of some great business leaders on both the supplier and the airlines sides, continuous improvement in reliability, durability, and performance have made air travel safe and economical for passengers around the globe.

GE Aircraft Engines, although the first to build a jet engine in the United States in 1941 and a primary engine supplier to the U.S. Department of Defense since then, did not get deeply involved in the commercial business until the late 1960s. GE had many technological firsts in this period, but probably the biggest was being the first company to run a high-bypass engine. This not only made it possible to build big military transport planes such as the C5A, but also showed the world such a design was possible and resulted in Pratt & Whitney and Rolls Royce building similar engines. The advent of a high-bypass engine, with its low fuel consumption, low noise and emissions, together with its high reliability, made possible the B-747, DC-10, L-1011 and Airbus airplanes that changed the character of the commercial aviation industry.

Also, because of the large investment costs of building such engines, GE looked into building partnerships around the world, some of which Bob Garvin, together with many others, helped develop. Our big breakthrough in this area was the partnership between GE, SNECMA, and MTU to build engines for Airbus, which not only got SNECMA and MTU into the commercial business, but permitted GE to be the supplier on the first wide-body twin, the A-300 with the CF6-50 engine. This resulted in our development with SNECMA of the CFM56 family of engines, which has become the model for international cooperation. More than 10,000 of these engines have been built for many different aircraft.

Of course, we had earlier successful licensing agreements with German, Italian, Japanese, Swedish, and British industries when they built the T64, T58, F404, and J79 engines for various military aircraft and helicopter systems licensed and sold around the world. As a result of many of these programs, Ray Small, Dick Smith, Fred Brown, Ed Bavaria, Bob Garvin, and several other GE people built long-lasting relationships with their counterparts in these countries who developed a high regard for how GE worked with them. All of this laid the groundwork for collaboration and partnerships on many of GE's commercial engines such as CF6, CT7, CF34 and GE90.

The engine business was an accumulation of the worldwide technical, personal and company relationships that made our worldwide leadership possible. I was proud to have been part of this and to have had Bob Garvin as part of our GE team.

Brian H. Rowe
September 1998

Introduction

Commercial aircraft engine manufacturing is the quintessential producer goods industry. A vast amount of capital is required to enter it and, once in, to stay competitive. Of the many entrants only a handful of manufacturers survive and they serve a concentrated market of a few hundred customers, many of them government-owned. Three firms, two American and one British, dominate the engine industry. As has been the case with much advanced technology, military applications have led the way.

The Soviet aircraft engine industry, like that in the United States, was set up to develop engines for military aircraft. During the Cold War, Americans and Russians saw each other as the enemy and had very little contact until *perestroika*. In 1990, I led a group from GE Aircraft Engines to a propulsion conference in Moscow. As part of our exhibit, we held a technology symposium where, along with a good lunch, our engineers could present details of GE products and take questions from the audience. About 300 Soviet scientists, engineers, academics, and administrators came to the conference, no doubt accompanied by few dozen technical intelligence specialists from the Soviet air force. Keep in mind that they represented the only national industry with a range of aircraft engines as broad and advanced as that of the United States. They listened to our technical sales pitches and asked many questions. At the last question, the hall fell silent: "Where does the money come from for new programs?"

They had lived all their professional lives with government funding and it was already clear that there would be no more of that. Even if in 1990 they understood profits and retained earnings, they had none to fall back on. They did have a vague hope that the West might come in and give them money to replace what they could no longer expect from their own government. It became clear to them quickly that this was not going to happen.

In 1994, I took part in the study of a new airliner designed by the Myasichchev Design Bureau and the Indian National Aerospace Laboratory. My client was a banker who was considering investing in the program and asked me about the realism of the Russian and Indian market forecasts and cost estimates. I told him that the manufacturer of a military aircraft can be reasonably sure that it will recover its costs. (Cynics in the United States have said that the higher the cost, the higher the contract price.) For the manufacturer of civil aircraft, the world is quite different: price is set by competition in the market; profit depends on low costs. It seemed to me that neither the Russians nor the Indians had a good understanding of their costs.

In the autumn of 1995, GE Aircraft Engines entered into a contract with the U.S. Department of Commerce to train a group of Russian aerospace engineers and managers. With the end of the Cold War, Russian orders for military equipment had vanished and government funding for development dried up. The Russian industry needed to adapt itself to a world where civil markets and commercial considerations determine survival. To the Russians, conversion meant selling their military products to non-Russian customers, and the U.S. government was trying to show them another way. One objective of the GE training course was to demonstrate how GE had made the conversion and become the world's largest producer of commercial aircraft engines.

GE asked me to speak on the subject. I spent more than 40 years in GE Aircraft Engines, had worked on many of its international programs, and had been responsible for developing the market for GE in Russia. To prepare for the Russians, I interviewed many of the people who had made the key decisions over the years. My notes were too interesting to set aside after the lecture. It would be a pity to lose their accounts of what happened, even when these individuals did not agree among themselves. Besides, memory fades. This book expands on the notes for the lecture. To give them context, I have added a history of the market and GE's place in it, along with my account of how GE Aircraft Engines converted itself from military to commercial markets. Because I was deeply involved, I have given a lot of space to GE's market development effort in China and in the USSR (later in Russia) to illustrate the challenges of market development in countries whose economic yardsticks are radically different from ours, and whose large military industry automatically becomes the chief competitor. Finally, I draw conclusions for potential investors, the Russians included.

The opinions expressed are mine. If there are errors of fact, they are also mine. I am grateful for the kindness of key present and retired executives of GE Aircraft Engines and of Airbus Industrie who shared their memories with me, among them:

A. P. Adamson	L. Kapor
R. Beteille	J. N. Krebs
F. I. Brown Jr.	G. Neumann
N. Burgess	J. S. Parker
Walter Chang	Richard Ridge
F. Ehrich	B. H. Rowe
Eric Falk	R. B. Smith
B. J. Gordon Jr.	R. C. Turnbull
M. C. Hemsworth	W. E. Van Duyne Jr.
B. C. Hope Jr.	Edward Woll

I am also indebted to the GE publication *Eight Decades of Progress* for key dates and facts and to my editor, Lois Malone, who straightened out the kinks in my text with the utmost kindness.

R. V. Garvin
February 1999

CHAPTER 1

How GE Got into the Jet Engine Business

The First Generation: 1903 to 1945

POWERED flight began in 1903 with the Wright brothers, who designed and built their own engine. In the subsequent enthusiasm for aviation, many aircraft companies were formed in Europe, and some of the most advanced aircraft engines were designed and produced in France. World War I led to large-scale engine production in Britain, France, and Germany, far greater than any such activity in the United States at the time. At the end of the war, the primary American engine was the V-12 Liberty, which had just reached production, and for a long time after that the existence of many war surplus Liberty engines put an end to any thoughts of aircraft engine development in the United States.

During the next 20 years, the aircraft engine industry in the United States consisted primarily of two firms: Curtiss–Wright, which later became Wright Aeronautical; and Pratt & Whitney, formed in 1925 by several engineers and executives who split off from Wright Aeronautical. Each firm designed and built a series of successful air-cooled radial engines for both military and commercial aircraft. The Allison division of General Motors was a smaller operation, developing specialized V-12 liquid-cooled engines for the U.S. Army Air Corps. Other smaller companies such as Continental Motors developed smaller engines, primarily air-cooled. All of these engines were of the reciprocating piston type, similar to those in automobiles, but turning propellers (with the exception of the French Gnome & Rhone engine of World War I; in this radial air-cooled piston engine, the propeller was fastened to the crankcase and both rotated as a unit, hence the name "rotary engine").

In 1878, with financing arranged by J. P. Morgan, Thomas Edison formed the General Electric Company to make and sell machines to generate and distribute electricity (such as electrical generators, steam turbines, transformers, switchgear, and copper cable) and products to use it (such as electric motors, light bulbs, and appliances).

GE entered the aircraft engine industry through a side door, as it were, by designing an accessory for piston engines, the turbosupercharger. This came about because one man was able to turn technical conviction into a real product, with the backing of GE's large technical and financial resources. Dr. Sanford E. Moss became convinced of the future of gas turbines while he was still an undergraduate at Cornell University in New York. He continued his research when he joined GE. There he developed the turbosupercharger and struggled to convince the users of aircraft that they needed such a device to improve the performance of their aircraft. Years later, this would lead the company into jet engines.

Moss joined the GE Turbine Research Department in 1906. His research team was a part of the Steam Turbine Department, which manufactured a line of centrifugal air compressors for blast furnaces and the chemical industry.

Air compressors can take several mechanical forms. For large air flows, the high-speed rotary compressor is preferred. It can be *centrifugal*, accelerating air by driving it with centrifugal force from the inlet at the hub of an impeller wheel to the rim, where the kinetic energy is converted to pressure. The compressor can also be *axial*, using several stages of compressor disks whose rims carry small blades similar to those of a small propeller.

At that time, industrial compressors were driven by steam turbines or powerful electric motors. Moss was convinced that hot combustion gases could be used to rotate a turbine, which in turn could drive the air compressor, and he pursued experiments to this effect at Cornell University, but because there were no materials capable of withstanding high stress at hot gas temperatures and because the efficiency of the compressors and turbines he used was too low to permit a self-sustaining gas turbine cycle, he achieved limited success.

Along the same lines as the compressor drive, in a turbosupercharger the exhaust gases of a piston engine rotate a turbine, which in turn drives a rotating centrifugal impeller. The impeller compresses air, which flows into the cylinders of the piston engine during the induction stroke. Because the air is compressed, a greater mass of it flows into the cylinder than in a normally aspirated engine. Add fuel in proportion and the force exerted on the piston during the power stroke is increased, resulting in greater power output with greater fuel flow or reduced fuel flow for any given power output.

The Swiss engineering company Brown–Boveri designed diesel engines and boilers for powerplants and ships. In 1911, Brown–Boveri experimented with a turbosupercharger for a diesel engine. In steam boilers, hot combustion gas turns water into steam. Brown–Boveri used a form of turbocharger for its "Velox" boiler: the hot exhaust gases leaving the boiler's combustion zone drove a turbine, which in turn drove a compressor supplying combustion air under pressure to the boiler.

In 1917, U.S. President Wilson set up the National Advisory Committee on Aeronautics (NACA), led by Professor William F. Durand of Cornell University. A French engineer, Auguste Rateau, proposed that turbosuperchargers be used to boost the power of aircraft engines at higher altitudes, where the air is thinner and a conventional engine loses power. Durand had supervised the Moss experiments at Cornell University and induced NACA to invite both GE and Rateau to develop working prototypes of a turbosupercharger.

A gasoline piston aircraft engine typically has an exhaust gas temperature higher than that of a typical diesel engine, and that poses difficult materials challenges to the designer of a turbosupercharger. Moss was able to draw on GE's experience in the design of high-rotating-speed steam turbines and, in the metallurgy of ductile tungsten and tungsten alloys used for its light bulb filaments and x-ray targets, to deal with the stresses in the turbine wheel of a turbosupercharger. His turbosupercharger was demonstrated successfully in 1918 on a Liberty V-12 aircraft engine. The unsupercharged Liberty delivered 350 hp at sea level at the U.S. Army's McCook Field in Dayton, Ohio, and was then run again in a test bench mounted on a truck atop Pikes Peak in Colorado, 14,109 ft above sea level (Figure 1). There, the output of the unsupercharged Liberty engine was 230 hp; with Moss' turbosupercharger, the engine delivered 356 hp, 6 hp more than at sea level. This triumphant demonstration was followed by dramatic high-altitude flights in 1920 by the U.S. Army Air Corps, in which experimental turbosupercharged aircraft attained altitudes of 33,000 and 40,800 ft above sea level, far higher than any altitude attainable by aircraft with unsupercharged engines.

Fig. 1 Starting the Liberty engine on the top of Pike's Peak, 1918. The turbosupercharger is mounted above and behind the propeller. (GE negative)

At the time, U.S. aircraft and engine firms, advised by pioneer aviators such as Charles Lindbergh, believed that air-cooled radial engines were lighter, simpler, and more reliable than liquid-cooled engines. Radial engines do not lend themselves to easy installation of a turbosupercharger, for which hot exhaust gas must be taken from each cylinder into a common manifold that feeds the turbine. To avoid the problems posed by such hot exhaust gas collector manifolds, Wright Aeronautical Corporation and Pratt & Whitney concentrated on integral superchargers, their impellers driven directly by the engine crankshaft. GE designed and built the centrifugal compressors for their mechanically driven superchargers.

But the U.S. Army Air Corps had not abandoned its interest in turbosuperchargers. In 1921, Brigadier General William E. "Billy" Mitchell demonstrated that his Martin MB-1 bombers with turbosupercharged engines could carry a bomb load to an altitude of 15,000 ft, higher than the range of anti-aircraft guns of the day. To prove that bombers could destroy armored warships while turbosupercharged engines allowed them to remain relatively invulnerable to the ships' defenses, he arranged to have his bombers sink two old German battleships moored off the U.S. coast. His strident advocacy of strategic airpower did not endear him to the U.S. Navy or the U.S. War Department. Some years later, the Japanese navy would perform similar demonstrations at Pearl Harbor and off the coast of what was then Malaya.

The U.S. Army continued to support GE's development of the turbosupercharger for the tactical benefit available from flight at higher altitudes. The problems of exhaust manifolds were finally solved and, in the mid-1930s, the first Boeing B-17 and later the Consolidated–Vultee B-24 bombers were designed to use GE turbosuperchargers on their radial engines. The Lockheed P-38 and Republic P-47 fighters also were designed with turbochargers, the former on its Allison V-1710 V-12 engine, the latter on its Pratt & Whitney R-2800 radial engine. During World War II, GE supplied some 300,000 turbosuperchargers for most of the American fighter and bomber engines, including the Wright R-3350 engines on the Boeing B-29 and the Pratt & Whitney R-4360 on its postwar successor, the Boeing B-50.

From 1948 to 1949, Northwest Orient Airlines, British Overseas Airways Corporation (BOAC), Pan American Airways, and several other airlines began to fly the Boeing Stratocruiser airliner on transoceanic routes. The Stratocruiser used the wing and engines of the B-50 heavy bomber. Its Pratt & Whitney R-4360 engines were equipped with GE turbochargers. This was GE's first experience in supporting such a product in airline service, an element of the business we will discuss in some detail in a later chapter. There were operational problems with the turbochargers, but these were not half as severe as the problems with the engine itself, a complicated radial with 28 cylinders arranged in four rows of 7. As it happened, the Stratocruiser was superseded quickly in airline service by more efficient airliners such as the Douglas DC-6 and DC-7 and the Lockheed Constellation, which also used versions of GE turbochargers on their Pratt & Whitney R-2800 and Wright Aeronautical R-3350 engines.

Precisely at this time, the U.S. Air Force and U.S. Navy began converting to jet power for their bombers and fighters, and it became clear to GE that the military would not be buying any more turbosuperchargers. GE's experience had been overwhelmingly with turbosuperchargers for military aircraft and only in a small way for commercial aircraft. GE decided to build a number of units ahead of airline demand and then to stop manufacturing turbosuperchargers. GE proposed to the airlines that they buy and hold these units until they needed them. Years later, when GE was trying to sell jet engines to Northwest, the airline would choose Pratt & Whitney engines for its DC-10. Northwest had had good experience with Pratt & Whitney engines and saw no strong reason to change to a new supplier. But it also pointedly reminded GE that it had not supported its turbosuperchargers.

Turbosuperchargers contain the two central elements of a gas turbine, a rotary compressor and a turbine to drive it. As an outgrowth of its work on turbosuperchargers, GE had been experimenting with various forms of gas turbine engines for the U.S. Army and U.S. Navy, as were the other two American steam turbine producers, Westinghouse and Allis–Chalmers. One of GE's developments was the TG100, a gas turbine driving an aircraft propeller. Unlike the GE turbosupercharger, it used an axial compressor, smaller in frontal area for a given airflow and pressure rise than the centrifugal compressor of the turbosupercharger, but with a turbine wheel very similar to that of the turbocharger.

Aircraft gas turbine development had been carried on since the early 1930s in England and Germany, with other experiments in Italy and Japan. The first successful gas turbine-powered flight, by a reaction jet pushing gas out of a nozzle in the back rather than by a propeller in front, was made in Germany in August 1939 on a Heinkel airplane powered by an engine developed by Hans von Ohain. In September, Germany invaded Poland and Britain declared war on the Germans. In Britain, a Gloster airplane powered by an engine developed by Frank Whittle flew on May 15, 1941. In September 1941, General H. H. "Hap" Arnold, the deputy chief of staff of the U.S. Army Air Corps, directed GE to build the British Whittle engine. Physically, the Whittle engine was similar to the GE turbocharger; both had centrifugal compressors and axial turbines, and comparable diameters. Arnold chose GE because he wanted the American aircraft engine manufacturers to concentrate on production of their piston engines. He also had some concern that they might not be enthusiastic about a revolutionary engine type that would eventually turn conventional aircraft propulsion practice on its ear.

In great secrecy, GE built an engine from Whittle's drawings and tested it successfully in April 1942. Versions of the engine flew for the first time in the Bell P-59 in October 1942. GE immediately began further development of the engine to attain greater thrust and stability and to substitute American structural alloys proven in its turbosuperchargers for those used by Whittle in England. In 1943, GE began the design of the I-40 engine, four times as powerful as the first Whittle engine. With the U.S. Air Force type designation J33, the I-40 went into production for the Lockheed P-80 Shooting Star, America's first operational jet fighter. The J33 was built in quantity by the Chevrolet and Allison divisions of General Motors.

Jet Power and the Cold War

World War II was over in 1945, with Germany and Japan defeated, but political tensions continued at a high level. The former Soviet

Union had extended its hegemony over the European countries on its western border, including the zone of Germany it occupied. None were allowed to join the European economic recovery program under the U.S. Marshall Plan. In the East, North Korea set itself up as a separate country. Chinese communists were successfully driving Chiang Kai Shek and his Kuomintang party, with whom the United States had been allied in the war against Japan, out of China. Clearly there would be a long political and military competition between the United States and the former Soviet Union, each aided by its bloc of allies. The United States had rapidly demobilized its army, but its navy and newly independent air force began to plan their force structures and equipment needs around the objectives of keeping a putative Soviet invasion out of western Europe and containing any Asian communist revolution inside China. For both services, that meant developing a new generation of jet-powered warplanes.

In 1945, GE began the development of the J47 turbojet engine for the North American Aviation F-86 fighter and the Boeing B-47 bomber, the F-86 using a single J47 engine, the B-47 six. The Korean War broke out in 1950, and the F-86 became the frontline fighter for the U.S. Air Force when North Korea deployed Russian-built MiG 15 fighters, which outclassed all of the American fighters stationed in Japan, both the propeller-driven F-51 Mustangs and jet P-80 Shooting Stars. The appearance of the MiG 15 made clear that the era of propeller-driven fighters and bombers was over. By necessity, the B-47 became the frontline strategic bomber for the U.S. Air Force, and the F-86 the frontline fighter for the U.S. Air Force and U.S. Navy. Both services began high-rate procurement of jet aircraft and J47 engines. At one stroke, GE absolutely dominated the U.S. market for military jet engines, delivering more than 33,000 J47s while another 5000 were built under GE license by Studebaker, Packard, FIAT in Italy, and Ishikawajima–Harima in Japan. The peak production rate during the Korean War was more than 1000 engines per month. GE made only about 15% of their parts, procuring others from a huge network of suppliers.

The traditional engine makers, Wright Aeronautical and Pratt & Whitney, at once set about to regain their leadership of the industry with new gas turbine engines. Both elected to get started by building British engines under license for the U.S. Air Force and U.S. Navy while they developed their own designs. Wright Aeronautical built the Armstrong–Siddeley Sapphire, designated J65 in the United States, for the F-84 fighter and B-57 Canberra light bomber, itself also a British design built by the United States by Martin; the J65 was also used in the Douglas A4D attack aircraft and in the first few Lockheed F-104 Starfighters. With its advanced axial flow compressor, the J65 had better performance than the J47. Pratt & Whitney took a license from Rolls–Royce to build the J46 Nene and J48 Tay, two rugged engines with centrifugal compressors that were used in early Grumman jet fighters such as the F9F Panther. Pratt & Whitney also began the development of an advanced engine of its own design, the J57, but Wright Aeronautical's corporate management would not approve similar work.[1,2]

In 1953, with the Korean War coming to an end and production of the J47 facing an inevitable decline, GE formed a corporate committee to study the strategic future of the aircraft engine business. Pratt & Whitney was the American leader in the commercial piston engine market, and its successful new military J57 turbojet engine promised to allow Pratt & Whitney to extend that leadership into the market for commercial jet engines. The GE committee noted that Wright Aeronautical, one of the three major producers of commercial aircraft engines (the third was Rolls–Royce), was beginning to lose market share and strength. Its chief product was the complicated R-3350 TC18 turbocompound engine, used in the Douglas DC-7 and Lockheed Constellation. The airlines experienced many problems with this engine and complained that Wright Aeronautical's management displayed inadequate responsiveness. Even more fundamental, Wright Aeronautical's gas turbine engine experience was limited to the J65. Wright Aeronautical's management expressed interest in developing the Bristol Olympus turbojet into the military J67 and into a commercial version (the Olympus, in its final form, became the engine for the Concorde SST, with an afterburner

---

[1]"The Has-Been," *Forbes*, October 1, 1996.
[2]Eltscher, L., "Curtiss–Wright: Greatness and Decline," *Journal of the American Aviation Historical Society*, Winter 1994.

developed by SNECMA in France). But the Curtiss–Wright board of directors withheld its approval, despite the expectation that, sooner or later, the airlines would demand jet-powered aircraft as the military had.

The other two commercial engine manufacturers, Pratt & Whitney and Rolls–Royce, made their reputation with reciprocating engines. As far as the GE committee knew, GE's gas turbine technology was the equal of theirs and more advanced in some respects. Airlines in the United States, dissatisfied with Wright Aeronautical, would welcome another dependable competitor to Pratt & Whitney. However, it was clear to the committee that, although it had been a supplier of electrical equipment and turbosuperchargers to the airlines, GE had no track record as a commercial engine supplier. The committee recommended that GE pursue three objectives: the civil as well as military jet engine markets, the development of a new high-performance Mach 2 fighter engine for the U.S. military services, and the development of small aircraft gas turbines for helicopters. The committee's chairman was Jack S. Parker, who had joined GE's Aircraft Nuclear Propulsion project in 1950 after successful production at the Todd shipyard in Oregon. After the committee made its recommendations, Parker was put in charge of the new small aircraft gas turbine department he had recommended. He retired from GE in 1979 as vice chairman responsible for all high-technology businesses.

In fact, the situation was not as straightforward as GE's confident assessment of its own technology might indicate. Here it is necessary to make a digression into the history of the military jet engines whose design provided the technical basis for the first generation of commercial engines.

The performance of the MiG 15 in Korea showed U.S. aircraft designers that they could not rest on their laurels. Pratt & Whitney and GE had both started design work on the next generation of high thrust engines for the U.S. Air Force in the early 1950s. Such engines needed a higher compressor pressure ratio than the 5:1 of the J47. GE took a very conservative approach, scaling up the airflow and increasing the compressor pressure ratio to 8:1 for the new engine. To attain stable airflow at this higher pressure ratio, GE designed inlet guide vanes adjustable from relatively closed at the start to wide open at higher compressor rotor speeds. Rolls–Royce adopted the same solution for its contemporaneous Avon engine. GE's designers had concerns about pressure ratios higher than 8. Very little was known about sealing airflow at such high pressures, and even less about the aerodynamic stability of the compressor.

GE designed two engines with this approach, the larger of which, the J53, never got beyond the prototype stage. The smaller, the J73 (Figure 2), was developed for the North American Aviation F-86H fighter, a similarly conservative scaling up of the F-86. This had a maximum gross takeoff weight of more than 20,000 lb, a single-engine fighter weighing as much as a DC-3 airliner with 21 passenger seats! Aircraft and engine worked as advertised. Both were sturdy, heavy, and dependable, and GE delivered a total of 891 J73 engines—less than one month's peak production of the J47 during the Korean War. But in service the F-86H was rapidly supplanted by a superior and more versatile aircraft, the North American Aviation F-100 with the Pratt & Whitney J57 engine. Unlike the F-86H, the F-100 was capable of supersonic flight and rapidly became the first-line fighter of the U.S. Air Force and allied air forces, as the F-86 had been before.

Fig. 2   J. N. Krebs (left) with a J73 engine. (GE negative)

Pratt & Whitney made the transition from large piston engines to jet engines by building Rolls–Royce jet engines under license while it developed its own gas turbine. To be successful, Pratt & Whitney's own development had to represent a significant advance over the J47. Like GE with the J73, Pratt & Whitney selected higher airflow for its new engine, the J57, but took a much bolder step with the technology of the compressor. To get the thrust and specific fuel consumption required for a transsonic fighter, Pratt & Whitney selected a pressure ratio of 12:1. To achieve stable compressor airflow, Pratt & Whitney divided the compressor into two parts, one behind the other, each driven by its own turbine running at its own speed. These mechanical and aerodynamic concepts were revolutionary and delivered performance far better than that of the J47 or indeed the J73. The J57 became an enormous success, gaining for Pratt & Whitney the same dominance of the American military market held by GE's J47 a generation earlier. The J57 engine was installed in the F-100 and other widely used supersonic fighters such as the F-101 and F-102. Without its afterburner, it was used in the subsonic Boeing B-52 bomber and the Boeing KC-135 tanker.

The bomber and tanker installations were prophetic: both can be thought of as transport aircraft of a specialized type, not-so-distant relatives of airliners. Pratt & Whitney certified a civil version of the J57 for commercial service as the JT3. Boeing installed the engine in the 707, followed by Douglas in the DC-8. With the J57 and its JT3 commercial version, Pratt & Whitney had convincingly displaced GE as the market leader.

By 1952, U.S. Air Force planners had already identified the need for a generation of fighters and bombers to follow the F-100 and B-52, aircraft capable of Mach 2 supersonic performance. GE was determined to regain the military market with an engine design delivering the necessary thrust, lighter in weight than the J57 but with the same good fuel economy. Such an engine would also require a pressure ratio of 12:1. There were two fundamentally different design approaches to stable engine operation at higher pressure ratios. One was the dual rotor approach used successfully by Pratt & Whitney's J57. The other kept the single compressor rotor but made the stator vanes adjustable, so that they were more nearly closed for starting and at low compressor speeds and wide open at maximum compressor speed. The single rotor needs fewer bearings and seals than the dual rotor and weighs less; against that, the adjustment mechanism for the vanes is complicated and one could expect leakage of high-pressure air around the base of the adjustable vanes. Compressed air is hot, and the airframe designer needs to avoid hot air inside the engine bay.

There are pluses and minuses to each design approach and, in 1952, there were few facts on which to base a choice. GE set up two separate teams to lay out preliminary designs and analyze their performance and cost. The variable stator engine team was headed by Gerhard Neumann, who had earlier led a small team to design and test an experimental compressor with variable stator vanes. The dual rotor engine team was led by Dr. Chapman Walker, one of GE's

Fig. 3 Gerhard Neumann and Neil Burgess with the J79 Engine, for which they won the Collier Trophy. (GE negative)

veteran designers. In October 1952, the two teams compared their results. GE decided to proceed with the variable stator concept and, in early 1953, Neumann headed the team to design and build a prototype engine (Figure 3). The outcome was the J79 engine, combining high thrust with good fuel economy and very light weight. The U.S. Air Force picked this engine for the Convair B-58 Hustler and the Lockheed F-104 Starfighter, both Mach 2 combat aircraft. Later versions of the J79 were picked by the U.S. Navy for the F4H Phantom, which became the main fighter for both U.S. services and for the air forces of Israel, Germany, Japan, Turkey, and Korea. (Britain also adopted the Phantom for the Royal Air Force but replaced the J79 with the Rolls–Royce Spey engine.)

Military Technology Flows into Commercial Jets

While struggles for the huge American military market were going on, the commercial jet age began in a very modest way. BOAC began jet service in 1954 with the medium-range DeHavilland Comet 1, powered by four DeHavilland Ghost engines roughly comparable in thrust to the GE I-40. The Comet 1 had to be withdrawn from service after a series of catastrophic accidents caused by fatigue failure of the fuselage, a problem corrected in the Comet 4, powered by four Rolls–Royce Avon engines. The British Overseas Airways Corporation began transatlantic service with the Comet 4 in October 1957. But even this version of the Comet could carry only 90 passengers and had limited range. The problem of limited range of jet aircraft was not new; early jet bombers were also limited in range. The first B-45 squadron stationed in England had just enough range to reach targets in the former Soviet Union but not enough to return. (Presumably, the flight crews were aware of this.) The U.S. Air Force Strategic Air Command needed an in-flight refueling system that would give jet bombers the range to fly intercontinental missions. While the first B-47s in Strategic Air Command had been more or less adequately supported by piston-engined tankers, the U.S. Air Force needed more capable aerial tankers to support the new fleet of B-52 strategic bombers. In July 1954, Boeing had built and flown a prototype jet transport, the

Model 367-80, with four Pratt & Whitney J57 engines. Based on the Model 367-80 transport prototype, Boeing designed the KC-135 jet-powered tanker with the same J57 engine already in service in the F-100 fighter and the Boeing B-52 bomber. The U.S. Air Force ordered more than 700 of these tankers.

It was a small step from the tanker to a commercial airliner. Boeing launched the design of the 707 and soon Douglas followed with the DC-8. These first American jet airliners were larger than the Comet 4 and had longer range, features that appealed to airlines flying across the Atlantic or Pacific Oceans. Both aircraft used the civil version of Pratt & Whitney's military J57, the JT3. Pan American Airlines began transatlantic service with the 707 in October 1958 and other airlines followed, slowly but surely pushing the piston-engined Douglas DC-7s and Lockheed L-1049 Constellations into service on shorter routes.

In 1953 Jack Parker headed GE's committee on the industry's future; in 1955, he took over management of the GE Aircraft Gas Turbine Division. Concluding that the airlines' perception of GE had to be changed, he asked the aviation consultant Dixon Speas to arrange meetings with airlines in Europe and the United States—not with their chief executives but, because it was going to take some time for GE to get ready, with the men who would replace them when they retired a few years hence. Parker, Speas, and Neil Burgess, the J79 project manager, visited the major airlines in the United States and Europe to discuss GE's plans in anticipation of the day when GE would enter the commercial engine business. They spent a month meeting and getting to know the men who would later be in charge of American, Delta, United, KLM, Swissair, and Scandinavian Airlines System.

It was clear that airlines flying routes shorter than those across the Atlantic and Pacific Oceans would also soon want to replace their piston-engined airliners with jet transports. Major domestic American carriers such as Trans World Airlines (TWA), United Airlines, and American Airlines were, in fact, interested in an airplane smaller than the 707 or DC-8 for medium-range routes. Convair, which was building a successful 44-seat twin piston engine airliner, saw a market niche for a new jet, and Convair asked Neil Burgess whether the

J79, which had powered its B-58 bomber, could be turned into an engine for a new airliner, the Convair 880. Burgess sketched a J79 minus its afterburner, added a thrust reverser and sound suppressor, and gave an on-the-spot price estimate of $125,000 per engine. He telephoned Jack Parker to get his approval and find out who owned the patent rights for the J79's variable stator mechanism. Such rights would not affect the purchase of a military engine by the U.S. government but could be a roadblock in the case of a commercial sale to an airline. Bill Gunston in his history of Rolls-Royce suggests that Rolls-Royce tested a compressor with variable stator vanes in the late 1940s, but abandoned the idea because of its mechanical complexity. In any case, Burgess got his approval and was promptly appointed to manage the program to convert the military J79 into the commercial CJ805.

The Convair 880 was lighter and faster than the 707 and DC-8. The lightweight CJ805 turbojet contributed to its high performance. But the 880 also carried fewer passengers, and several airlines asked Convair for a larger version, which would need more thrust. Convair laid out the design for the larger 990, and GE developed a more powerful turbofan version of the CJ805, the CJ805-23B.

The Convair 990 was the first American jetliner with a turbofan engine to increase thrust and reduce specific fuel consumption and noise (the first practical front fan engine for commercial airliners was the Rolls–Royce Conway, used by BOAC in its versions of the Boeing 707, and by Alitalia in some DC-8s). When, in 1952, GE set an objective for a commercial engine capable of transatlantic service, Chapman Walker's design team built a demonstrator turbofan engine, a single-rotor front fan, with the fan mechanically coupled to the same turbine that drove the compressor, perhaps to design around the dual-rotor approach adopted by Pratt & Whitney. The experiment was not a success; the engine was difficult to start and operate. That experience and the impracticality of putting a fan in front of the CJ805 engine led GE to an ingenious solution for the Convair 990—a single stage aft fan, in fact a free turbine wheel behind the turbojet carrying fan blades at the tip of the turbine blades (Figure 4). There was no need to make great changes in the turbojet itself but unlike the front fan, the aft fan added no pressure to the

Fig. 4 Neil Burgess with a CJ805-23B display engine. The aft fan, on the right, is attached to the rear of the gas generator, behind the 3-stage turbine which drives the 17-stage compressor. (GE negative)

airflow through the core engine. The aft fan engine was not as easy to install in the aircraft as the front fan and weighed more for a given thrust, but, as the performance advantages of the CJ805-23B aft fan became clear, Pratt & Whitney, which had at first belittled the concept of the turbofan, changed its mind and added a fan at the *front* of the JT3. Aft and front fans have advantages and drawbacks; the airliner market quickly demonstrated its preference for the front fan.

Pratt & Whitney's choice of the dual rotor compressor over GE's single rotor with variable stator vanes was crucial. It was relatively simple to add a fan to the front of the J57 and JT3, coupling it mechanically with the low-pressure compressor, which the fan supercharged. Most of the major carriers selected the Boeing 707 and Douglas DC-8, first with Pratt & Whitney JT3 turbojet engines and later with the JT3D, the version with a front fan. The front fan versions also found an instant military application in the B-52H bomber and C-141 military transport.

Airline traffic had grown rapidly with the introduction of jet airliners. The fuselage of the Convair accommodated 110 to 149 seats, 5 abreast. Its seat–mile cost was higher than that of the 707, which had as many as 185 seats, or that of the DC-8 with as many as 259 (in both aircraft, 6 abreast). In 1956, Howard Hughes, the eccentric millionaire who owned TWA, ordered 30 Convair 880s and Delta Airlines ordered another 10. Eight more were ordered by airlines in Brazil and Argentina. Later, there were orders for the somewhat bigger 990 version with the CJ805-23B aft turbofan from American Airlines and Swissair. Japan Airlines, Garuda, and Cathay Pacific, all quality airlines, ordered a few Convairs. But a dispute between Howard Hughes and Convair delayed entry of the 880 into service, while the 707 and DC-8 were establishing a strong market position. GE enhanced its technical reputation with the installation engineering support it had given Convair to resolve the transsonic drag problem, which had delayed the Convair 990's entry into airline service. In the end, Convair sold only 102 airliners, enough to put its and GE's reputation on the line but too few to create an economic base for profitable spare parts sales. Pilots and passengers liked the aircraft, but it was too small to be competitive. The biggest problem may have been the lack of management focus at Convair, one of several divisions of General Dynamics, which seemed to lack a clear sense of direction and management support for its jet airliner business. By the time Convair conceded that it could get no more orders, it lost $600 million on this program, at that time the biggest single loss in American industrial history. With that, Convair gave up the airliner business entirely.

There was one other application for the CJ805-23B aft fan, equally unsuccessful. Aerospatiale, the airframe company owned by the French government, designed a twin jet airliner, the Caravelle, for short and medium routes. The Caravelle used the DeHavilland Comet's nose and cockpit and two Rolls–Royce Avon engines, the same engines used in the final version of the Comet. Several hundred Caravelles were built for service in Europe, and Capital Airlines also used some for a while in the United States, although these were soon displaced by the Douglas DC-9 and British Aerospace BAC1-11. In 1961, Aerospatiale suggested to Burgess that GE buy a Caravelle and install the CJ805-23 in place of the Avon to demonstrate the GE engine to European airlines. Burgess agreed. GE's flight test people at Edwards Air Force Base in California modified Caravelle No. 42, which made the rounds of potential European customers. Christened *Santa Maria*, the name of Christopher Columbus' flagship, this Caravelle had the distinction of being the first jet airliner to land in Yugoslavia. No airline bought a Caravelle powered by GE, but this began a long association between GE and the French aircraft industry.

GE lost $80 million gross on the CJ805 program, half that after taxes, a large amount of money at the time. The Convair 880 and 990 aircraft stayed in service for a number of years, with charter airlines after service with the major carriers. GE continued to support the CJ805 engines and, in fact, managed to break even with subsequent spare parts sales.

CJ805 Service and Product Support

What airlines expected in the way of technical publications, service engineering support, and availability of spare parts was clearly different from the practices of the U.S. military. In 1956, GE hired the former head of American Airlines' maintenance and engineering division in Tulsa, retired U.S. Air Force Major General John B. Montgomery, to run the production engine department. Montgomery told his inexperienced GE crew to visit American and Capital Airlines and find out what an airline expected from an engine supplier in the way of support. They listened to what the airlines had to say and decided to make it happen.

GE set up a completely separate product support organization for its CJ805 engines and hired the chief engineers of both airlines as well as a number of other airline people to help set up new systems—for technical publications and service bulletins, warranty administration, packaging, order service, tracking parts serial numbers, and warehousing in accordance with standards developed by the

Air Transport Association (ATA) and International Air Transport Association, the technical associations of airlines. These systems were different from those for the U.S. military, and GE's engineers helped the ATA write many of the new airline standards for jet engines.

The Wright Aeronautical R-3350 Turbo-Compound, the last of the big reciprocating engines, complex and fragile, with poor reliability and comparatively short overhaul intervals, was still in wide airline use. Piston engines in general require more maintenance than gas turbine engines, and the airlines had reluctantly learned to live with such problems. During its discussions with the airlines GE learned what they did not like about the product support provided by Wright Aeronautical, and even Pratt & Whitney. GE decided to organize itself to avoid the problems of the Turbo-Compound. The challenge was to turn that engine's minuses into pluses for the CJ805.

Walter Van Duyne, Wright Aeronautical's service engineering manager, was bearing the brunt of the Turbo-Compound's problems in airline service. He and his service engineers were doing their best to help the airlines survive, and he enjoyed a good personal reputation despite the engine's problems. However, Wright Aeronautical's management never gave its product support organization the resources required to fix the problems with the engine but continued to raise spare parts prices until the presidents of United and Braniff Airlines both complained that they would never buy another Wright Aeronautical engine. In effect, Wright Aeronautical liquidated the goodwill of the company; by saving money that it might have spent to improve the product, Wright Aeronautical abdicated any opportunity to sell another engine. Taking that lesson to heart, GE also hired Van Duyne in 1967 to manage its CJ805 customer service.

The military J79, designed in 1953, had barely gone into service when its commercial version, the CJ805, began to fly in the Convair 880, so that GE had very little operating experience with the design ahead of airline revenue service. The engine was light in weight and high in thrust, more fragile in some ways than the Pratt & Whitney JT3 to which it was compared in the maintenance shops of the airlines. GE learned quickly that its engine was not rugged enough for airline service, and costly to maintain. At the suggestion of United Airlines, GE leased a Douglas RB-66 bomber and equipped it with two prototype CJ805-3 engines to fly simulated airline routes at Edwards Air Force Base in California. The idea was that GE would accumulate some "fleet leader" experience on the engine before the first Convair 880 went into revenue service. GE hoped that following United Airline's suggestion would help the 880 win an order from the airline, but Boeing had no intention of allowing Convair to win a foothold and made a very aggressive proposal for a low-price version of the 707. United Airlines picked the Boeing airplane and its Pratt & Whitney JT3—the comfort factor of a known and trusted supplier against the unknown Convair/GE product and product support. (United Airlines, Boeing, and Pratt & Whitney had all been divisions of a conglomerate called United Aircraft until the Justice Department had forced its dissolution.)

In terms of sales, the CJ805 program was a small "country cousin" to GE's big and profitable military engine business and a considerable cost drag. Getting management attention and resources to improve the engine was not easy. But Jack Parker insisted: it was vital for GE's long-term objectives that the CJ805's problems be fixed. GE finally put a lot of resources into the program and managed to increase the durability of the engine considerably. John Montgomery even separated the product support people from any responsibility for spare parts sales, which was handled by a separate sales organization. He wanted GE's service engineers to make their decisions only on the basis of fixing problems and making the product do what it was supposed to do, untainted by any desire to make a profit on the sale of spare parts. Engineers and military commanders are fond of such lofty ideals. GE's credibility in product support was doubtlessly enhanced by this policy. However, service and spare parts are two essential and inseparable elements of product support, and GE later combined them as two functions under one general manager.

You can do all of the right things, but that will not necessarily count if you do not also "hug" the customer. In 1966 and 1967, during the competition for the supersonic transport (SST) engine, the Federal Aviation Administration asked U.S. airlines to state their preferences for the competing designs of aircraft and engines. Foreign airlines had no official vote but were also polled. GE spent a lot of effort asking the airlines what they wanted in an SST engine and wrote the design specifications for the GE4 SST engine to reflect the responses. The GE4 got most of the airline votes because of the technology of the design and GE's reputation for standing behind the CJ805 and resolving its problems. Even Japan Air Lines did not vote against GE. VAL had used the CJ805 in Convair 880s for some of its north-south routes, and at that time had not been satisfied with GE's product support.[3]

Although the CJ805 program was not a commercial success, without it GE would not be in the airline engine business today. The company demonstrated that it was prepared to back up its product, listen to the airlines, and respond to their needs. The CJ805 program did not make a big profit but, in the end, it was a cheap ticket to the ballgame: GE discovered product support as a competitive tool.

---

[3]The events are discussed in more detail in Chapter 14.

Chapter 2

Business Jets

SOME time after 1956, James S. McDonnell, the chairman of McDonnell Aircraft, decided to enter the commercial aircraft business. Until then McDonnell had built only fighter planes. "Mr. Mac" had his engineers lay out the design of a corporate aircraft with four Bristol–Siddeley Orpheus turbojets and asked GE's Jack Parker and Gerhard Neumann whether GE could give him a better engine. GE had designed the J85 turbojet for several military applications, among them the GAM-72 decoy missile and the Northrop T-38 trainer and F-5 fighter. The engine was light and compact, with a good thrust-to-weight ratio, and was in volume production at GE's jet engine factory in Lynn, Massachusetts. Neumann proposed to add an aft fan to the J85 similar in concept to that of the CJ805-23B in the Convair 990. With a bypass ratio of 2, this CF700 engine had a takeoff thrust of 4200 lb and low fuel consumption, a good match for the McDonnell aircraft. (GE model designations use CF for commercial [turbo] fan, and CJ for commercial jet. This simple and logical system was not used for the first turbofan, the CJ805-23B, or for the most recent, the GE90.) Neumann put Jim Krebs, one of his engineers from the variable stator demonstrator program, in charge of developing and certifying the CF700. Mr. Mac discussed the marketing of the McDonnell 220 with Juan Trippe, the founder and chairman of Pan American Airways (PAA), who thought that PAA could own and operate a fleet of these executive aircraft for corporate users or lease the aircraft to them. But the deal collapsed, supposedly because Mr. Mac refused to make changes to the cabin interior proposed by Trippe. After 1,000 hours of testing, GE put the CF700 program on hold.

Meanwhile Aero Commander, a manufacturer of small aircraft in Dallas, Texas, was designing an executive jet aircraft around the Pratt & Whitney JT12 turbojet. GE told Aero Commander that by removing the aft fan from the CF700 it could create a simple turbojet of the right size and thrust for the Jet Commander. Because the new engine had less thrust than the CF700, GE felt that it should have a lower model number and called it the CJ610.

In 1959, the aviation pioneer Bill Lear Jr. was developing a new executive jet aircraft in Switzerland and needed an engine. The Learjet was to be pressurized and able to fly above turbulent weather. It would also be considerably faster any other executive aircraft. Lear visited GE's Lynn factory and was shown the J85 military turbojet, which was of the right size in its CJ610 version. Lear expected to sell hundreds of his aircraft; a low engine price was important for such a market and would have to be flat over the life of the program, unlike military prices that could be high when startup costs were high but would decline with additional production. GE developed an attractive price for the CJ610, much as Burgess had a few years before for the CJ805.

There was another even more attractive opportunity for the CJ610. North American Aviation, the designer of the F-86 fighter, had designed a small twin-engine utility transport for the U.S. Air Force, the T-39 Sabreliner. The U.S. Air Force wanted several hundred Sabreliners, and the CJ610 was a precise fit for it. Such aircraft are flown about 600 hours per year, as opposed to the typical 2000 to 4000 hours per year of an airliner, so that business jet engines can be expected to last a lot longer. Developing and certifying the civil CJ610 version of the J85 would be relatively simple, an attractive business incremental to the big J85 production expected for the F-5 fighter.

Gerhard Neumann had taken over management of GE's entire aircraft engine business when Montgomery left the company in 1961. The Sabreliner and Lear's proposal made sense to him. A U.S. Air Force order would justify the expense of the type certification and create the production base load; the Learjet would follow. In fact the Sabreliner did come first, and GE received an order for the engines from North American. Lear was a clever entrepreneur heading a small team of capable engineers, but it took him several years to give GE a firm order for engines and by then the price of the engine had risen. Lear was incensed, and for a while threatened to get J85 drawings from the U.S. Air Force at Wright Field to manufacture CJ610 engines himself to show GE how to make low-cost engines. Eventually, he placed an order for engines and began a long and successful association with GE.

During one of his trips to Europe, Neumann also proposed the CJ610 to the German airframe company Hamburger Flugzeugbau, which had hired Bill Lear's chief designer from the Swiss factory where the Learjet prototype was being built. Neumann also visited the French jet fighter manufacturer Avions Marcel Dassault, which was also designing a business jet, an aircraft bigger than the Learjet that would need a more powerful engine. Neumann proposed the CF700.

At a GE corporate staff meeting, Neumann briefed GE chairman Ralph Cordiner about these opportunities. Cordiner was intrigued. He had a cattle ranch in Florida next to one owned by Juan Trippe, to whom he described the Learjet. For the second time, Trippe jumped at the idea. There would be a lively demand for such aircraft in the United States, where distances between major cities are big enough to justify the purchase of an executive jet. American companies are generally more open than European companies to new concepts such as business jets. Pan American Airways intended to create a separate division to market and support such aircraft. A couple of weeks later, Trippe and his technical advisor, Charles Lindbergh, met with Cordiner and Neumann, and Trippe asked Neumann how many orders he would want to launch the CJ610 program. Neumann had not discussed a firm decision for a go-ahead with Cordiner and, with a sideways glance at his boss, said that he would proceed with an order for engines for 100 aircraft; Trippe offered to order engines for 200 aircraft and asked Neumann to work out a price and the delivery schedule so that they could get together with Lear. While he was doing this, Trippe decided to see whether other European airframe manufacturers could give him an aircraft with a cabin bigger and wider than that of the Learjet. A PAA technical team led by the airline's chief engineer, Frank Borger, began negotiations with DeHavilland, the designer of the world's first jet liner, the Comet, and asked GE to come along to negotiate an engine agreement in parallel. Ed Woll, the new manager of the Lynn plant, led the GE team, only to see PAA break off negotiations when DeHavilland failed to give it the price and performance PAA considered vital for the market.

The PAA team immediately flew from England to France to negotiate with Avions Marcel Dassault, for whose aircraft Neumann had proposed the CF700. Pan American Airways wanted something distinctive for the executive aircraft market; a turbofan engine met this criterion, and GE's CF700 was the only suitable powerplant. GE and Dassault entered into negotiations. There was little question that GE's engine would meet Dassault's requirements for payload, speed, and range; the French government was prepared to assist Dassault in funding aircraft development, tooling, and initial production. The major issues were engine price, and penalties in case GE failed to meet its deliveries. Neither company had much experience in this and, after difficult negotiations, they agreed that GE would offer a low introductory price for the initial block of engines and a cap to any penalties. In fact, there were questions about the drag of the engine pod and its inlet recovery, a measure of aerodynamic efficiency. GE and Dassault cooperated on finding a solution, with GE performing wind-tunnel tests at its own expense until the final design met the expectations of both companies. GE delivered engines to Dassault ahead of schedule, laying to rest any question of penalties.

The Dassault aircraft, the Mystère 20, met the PAA specifications and was quickly rechristened the Fanjet Falcon. Pan American Airways became the exclusive sales agent for the Falcon by ordering the first 200 with GE Aircraft CF700 engines. GE arranged to have the French company Hispano–Suiza manufacture the engine pod in France to GE's design, and to provide service and technical support for the flight test program. The Falcon business was very successful for PAA and Dassault. (Dassault was impressed enough by GE's performance to ask GE for J85 engines for its new Jaguar fighter. Ed

Woll made two engines available, but the French government insisted on the use of the Franco–British Turbomeca/Rolls–Royce Adour engine.) As GE's first big program in France, its success and the resulting large exports to the U.S. laid a foundation for subsequent GE commercial engine activities there. Meanwhile, certification of the United States. Air Force's T-39 Sabreliner and the Learjet had also gone ahead with the CJ610 engine. The Learjet was smaller and faster than the Falcon and carried a lower price. It quickly became popular and was sold in larger numbers than Bill Lear's original wildly optimistic forecasts. Aero Commander, having incorporated the CJ610 into its business jet, sold its design to Israel Aircraft Industries (IAI), which made and sold a good number as IAI Jet Commanders. In contrast, the Hamburger Flugzeugbau 320 Hansa was not a big success: only 13 aircraft were built and used by the German air force as "taxis" for cabinet ministers and generals. The Sabreliner filled a similar role in the U.S. Air Force, but in much larger numbers. Pratt & Whitney's JT12 engine found only one major application, a four-engine Lockheed executive aircraft called the Jetstar. (An even smaller market was found for a turboshaft variant of the JT12, used in a handful of Sikorsky Skycrane helicopters.) The Jetstar was expensive and sales were minimal. Because of GE's economies of scale from J85 production, the manufacturing cost of the CJ610 was much lower than that of the JT12. Pratt & Whitney faced the problem of supporting a small population of installed engines whose costs could not be reduced through increased production.

There are important differences between product support for business jet engines and airline engines. The typical business jet customer owns one or at most two aircraft (according to the *New York Times* [July 27, 1997], there are some 11,500 business jets and turboprops in the United States, owned by about 8,000 companies) and sees the executive aircraft as a fast long-range limousine. The typical business jet customer usually flies an average of 600 hours per year, has no maintenance shop, and does not want to carry any inventory of spare engines or spare parts. Such a customer needs to have service distributors available at convenient airports, carrying spare parts inventories and authorized and trained by the engine manufacturer to perform maintenance. The engine manufacturer needs to have a central overhaul facility, also at an airport, which can make available exchange engines and parts so that customers' aircraft can continue in service while overhaul is being performed. The manufacturer's service engineers need to visit the service distributors regularly to provide training and review warranty claims, and for spot checks with customers regarding their operating experience.

GE did all of that. The sheer size of its business jet program was an important difference from that of the CJ805. GE dominated the business with engines on more than 1000 aircraft in North America, a market position that gave GE a big base of engines over which to spread the fixed costs of product support. Although business jet engines do not consume many spare parts, there was adequate margin in the engine price, enough volume to make the service and overhaul business profitable, and the J85 production line in Lynn was a ready source of production hardware at low cost. The CJ610 service shop in Strother, Kansas, was much more successful than the CJ805 service shop in Ontario, California, simply because of its greater volume of work.

All in all, executive aircraft engines were a big business for GE. But small engines sell for less money than big engines, even when the profit margin is the same percentage of engine price. The business jet engine project always had to fight for management attention and resources against the projects for big engines. Management has to allocate finite resources to various projects according to their strategic priorities. Such resource constraints would soon lead GE to cede an important portion of the executive jet engine market to a new competitor.

In 1965, GE's advanced product planning operation had been thinking about an engine to follow the CF700 and CJ610. GE had just won a contract from the U.S. Air Force to develop the TF39, a large high-bypass turbofan engine, and was anxious to apply this revolutionary technology to a smaller business jet engine. The T64 turboshaft engine under development for the U.S. Navy offered a likely vehicle. With its efficient compressor, the power takeoff shaft driving a front high-bypass fan would yield excellent specific performance, better than that of the CF700. Lynn built a demonstrator "CTF64" of some 5000 lb takeoff thrust and calculated that detailed product design and certification would cost a mere $10 million in contrast with several hundred million to develop an entirely new engine. Nevertheless, Neumann and his advisors decided not to go ahead with this new turbofan. GE had its hands full with the military TF39, and the next step was going to be the development of a big commercial engine; Neumann was not prepared to divert resources to the certification of a new executive aircraft engine.

Garrett Turbine Engine Company was a successful manufacturer of small gas turbines used as starters for jet engines and portable auxiliary electric power units. If GE was not going to develop a successor to the CF700 and CJ610, Garrett could see an opportunity to take over a market niche that might not be vital for GE but would be very attractive to a company with different priorities. Based on its starter gas turbine, Garrett developed and certified the TFE731 series of turbofan engines and with them substantially displaced GE as the leading manufacturer of engines for executive aircraft.

Even as GE consciously conceded the business jet engine market to Garrett in 1965, it began to develop the TF34, a larger version of the CTF64, for the U.S. Navy S-3A patrol aircraft and the U.S. Air Force A-10 attack aircraft. The TF34 engine had excellent performance and was designed for a demanding 6-hour patrol mission from aircraft carriers. Before it went into service, GE had performed some 10,000 hours of engine testing, including two back-to-back 150-hour cyclic endurance tests on a single engine. Despite all of this preparation, the TF34 developed serious problems in its hot section as it went into fleet service. The design specification focused on the antisubmarine patrol mission and not on the training syllabus typical of peacetime service. Constant touch-and-go landings and takeoffs led to serious problems of thermal fatigue in the combustor, the turbine stator, and the turbine blades. Similar problems arose when the engine went into service in the U.S. Air Force's A-10 attack aircraft. GE quickly developed solutions to these problems, improvements that also made the engine suitable for commercial service. GE certified the commercial version as the CF34 in 1982, 17 years after handing over the market to Garrett by default. The CF34's first application was in the Canadair CL601 Challenger, a large and luxurious executive aircraft competing with the Dassault Falcon 900 and the

Gulfstream III. Garrett tried to match the CF34 with a revolutionary but complicated engine, the ATF3. A few of these were ordered by the U.S. Coast Guard for a maritime patrol version of the Dassault Falcon, but the Garrett engine never saw commercial service.

Canadair stretched the CL-601 executive aircraft and made it into a 50-seat regional airliner, also with the CF34. Operating conditions and product support requirements for the engine became the same as for engines in big airliners (happily, the U.S. Navy insisted on long parts life and high reliability for its patrol aircraft engines), except that typical regional airlines usually have smaller financial resources than major carriers. They require more sales financing than the larger carriers and pay their bills more slowly. Like business jet operators and unlike the major carriers, many regional airlines depend on the manufacturer and a few service distributors for maintenance support. In that sense, CF34 airline product support parallels that for executive aircraft in structure, with one significant difference: to corporate users, reliability is the most important criterion; to an airline, cost is equally important.

Although the CF34 was able to accommodate the growth of the CL-601 from an executive aircraft into a 50-seat airliner, a further stretch of the Canadair Regional Jet to 70 seats demands considerably more thrust. To protect its strong market position, GE is presently developing a more powerful version of the CF34 around a completely new compressor. Two Japanese engine companies have joined GE, each developing certain engine parts in the expectation of becoming the manufacturing source for them when the engine went into production.

Chapter 3

The Years of the Cold War

THE CJ805 commercial engine program was viewed by the industry as a failure, although that did not relieve GE of its responsibility to support the 416 installed engines in service with several airlines. GE's other commercial engine program, business jet engines, was solidly profitable but small in scope compared to the military programs at GE or the commercial and military programs at Pratt & Whitney. Pratt & Whitney had a big market with the military J57 engine, used in the F-100, F-101, and F-102 fighters and the B-52 bomber and KC-135 tanker. A scaled-up version, the J75, was used in the F-105 fighter–bomber and the F-106 interceptor. In the early 1960s, Pratt & Whitney developed a smaller dual-rotor engine for the U.S. Navy, the J52. This was to be used for the U.S. Navy's A-4 attack aircraft and the Hound Dog missile carried on some B-52s. These were all big production programs, with the U.S. government paying all of the bills plus a negotiated overhead and profit. Therefore, they exerted little pressure on Pratt & Whitney's own cash resources.

Pratt & Whitney certified the J57 as the JT3 commercial engine and, by adding a front fan, created the successful JT3D. With these two models, Pratt & Whitney began its domination of the commercial market as well. What had worked so well on the J57 also worked on the J52. Adding a front fan turned it into the commercial JT8D, the engine adopted by Boeing for the 727 and 737 and by Douglas for the DC-9. Like the JT3D before it, the JT8D became a phenomenal success for Pratt & Whitney, which built some 14,000 of them over a period of more than 20 years.

At the end of the 1950s, several competitions forced the GE designers to redirect their thinking from engines for supersonic fighters and bombers to engines for subsonic transports. The U.S. Air Force launched a competition for a new jet freighter, the Lockheed C-141, for which GE proposed an aft fan version of the J79, but Pratt & Whitney's front fan version of the J57, the TF33, was selected. GE then lost another competition to the Pratt & Whitney J52 for a subsonic U.S. Navy attack aircraft, the so-called Eagle Missile Carrier, which (unlike the C-141) never went into production. Finally, the U.S. Air Force again picked the Pratt & Whitney TF33 for the advanced B-52H bomber.

Soon after that, the Federal Aviation Administration (FAA) launched a competition for a supersonic transport. As usual, GE, Pratt & Whitney, and even Wright Aeronautical competed for the engine. Boeing, Lockheed, and the major airlines in the United States were invited to vote on their engine preference. Important airlines abroad were also consulted. Based on their votes and the FAA's technical evaluation, Boeing chose the GE engine, an afterburning turbojet with a variable stator compressor and an accessory gearbox cooled with fuel against the high ambient temperatures of flight at Mach 2.7; Lockheed picked the Pratt & Whitney design, a turbofan with afterburning in the bypass duct; the airlines and the FAA selected the Boeing/GE combination. Then the U.S. Congress began to grapple with the huge cost of developing and certifying the SST and its engine, and the problems of harmful exhaust emissions in the stratosphere, airport noise, and sonic booms, which collectively would lead to the termination of the program. However, GE finally won a competition for a commercial engine, even though the U.S. government was paying most of the bills. (Twenty-five years later, the U.S. government approved NASA's funding of a research program for Boeing, Pratt & Whitney, and GE to develop the technology required to solve the problems of the sonic boom and exhaust emissions. Japan also launched an SST research program and invited the two U.S. engine companies to participate with Rolls–Royce and SNECMA.)

In 1962, tired of losing competitions, GE set up an advanced product planning operation reporting directly to Gerhard Neumann, with a mandate to consider future markets and the products needed to win them. GE began to study dual-rotor front fan engines, the cycle that led to Pratt & Whitney's huge market share with the J57, its commercial JT3D version, and the subsequent JT8D. GE's first such design was an afterburning engine for the TFX, a new attack aircraft that U.S. Defense Secretary MacNamara intended be adopted by both the U.S. Air Force and U.S. Navy. The U.S. Air Force did adopt the TFX as the F-111; the U.S. Navy did not, but both services ended up using the same engine. The government picked a new Pratt & Whitney engine design over GE's for the F-111 and the U.S. Navy's F-14 fighter. The TF30 was also a dual-rotor front fan with afterburner, but Pratt & Whitney was deemed to have more experience with such designs than GE and therefore posed a lower risk. The Pratt & Whitney TF30 proved to be a disappointment in operation, subject to frequent compressor stalls in the F-111, and with marginal thrust for the U.S. Navy's new ship-board fighter. Because both aircraft depended on this engine, the U.S. government was forced to spend large sums of money with Pratt & Whitney to correct its problems. Years later, Pratt & Whitney's F100 turbofan engine also encountered severe problems in the F-15 fighter; once again, the U.S. Air Force spent large sums of money to rectify the problems.

After its bad experience with the TF30, the U.S. Air Force enlisted GE to develop an alternative to the troubled Pratt & Whitney F100 engine. Edward Woll, then general manager of the GE military engine operation, was convinced that the F101 bomber engine's core could serve as the basis of a fighter engine superior in performance to that of the Pratt & Whitney F100 and simpler in mechanical design. Such an engine would need a fan of lower bypass ratio and a different afterburner, easily scaled up from that of the F404. Woll used GE funds to build the "F101 Derivative Fighter Engine" prototype and, after successful ground and flight testing, persuaded the U.S. Air Force to support the program. The U.S. Air Force's objectives were to goad Pratt & Whitney into correcting its problems with the F100 and have an insurance policy in case Pratt & Whitney did not succeed. The F101DFE turned out to be so successful technically that the U.S. Air Force opened the F-16 production program to the GE engine as well as to the Pratt & Whitney F100. The GE engine proved to be price competitive, more powerful, and without the flight restrictions imposed by the F100's problems. The U.S. Air Force began to use the new GE F110 engine in its F-16s; the U.S. Navy finally had an engine powerful enough for its F-14s. This is discussed in more detail in Chapter 8.

In the wake of the unsuccessful Convair 880 and 990 programs and the loss of the TFX and C-141 engine competitions, GE's advanced product planning operation set itself a new task: to gain a competitive edge in subsonic turbofan engines, where the standard had been set by the Pratt & Whitney JT3D and JT8D engines, with compressor pressure ratios of about 16:1 and uncooled turbines. Parametric studies showed GE that a pressure ratio of 20:1, with a high-temperature air-cooled turbine, would give superior performance. The dual-rotor front fan pioneered by Pratt & Whitney was clearly superior to the aft fan concept used by GE in the CJ805-23; overall performance would become even better if GE variable stator technology were combined with the dual-rotor compressor.

Boeing proposed to the U.S. Air Force that a large transport capable of moving all the equipment of an army heavy division would greatly augment the strategic mobility of U.S. armed forces. The C-141 had just gone into use and could carry relatively dense cargo such as ammunition, but it was limited in the *volume* of cargo it could accept. Boeing visualized the CXX, a very large freighter capable of carrying tanks, with a maximum gross takeoff weight of some 800,000 lb. Such a huge aircraft would need a total installed thrust of some 150,000 lb for takeoff. For thrust of that order of magnitude, the B-52 bomber uses eight Pratt & Whitney TF33 engines (the smaller C-141 uses four). This could have been a feasible propulsion system for the CXX, but GE argued that such a revolutionary aircraft needed a new engine system—four advanced high-bypass turbofans with the latest technology of high pressure ratios and high turbine temperatures. The U.S. Air Force agreed. In 1964, GE and Pratt & Whitney competed for the engines for the new outsize strategic airlifter, designated the C-5A. The engines would have to be bigger, more powerful, and more efficient than any built to date, and they were going to be ordered under a novel form of contract, designed to avoid for the U.S. Department of Defense (DOD) the huge cost overruns of the F-111 and its Pratt & Whitney TF30. The contract would be a "total package procurement," one lump sum for development and production years into the future, shifting the risk of overruns from the DOD to the manufacturers.

GE saw this as an opportunity to define engine technology for the next 50 years, and took a giant gamble. The highest bypass ratio until then had been just under 3:1, in the CJ805-23 and the CF700; GE proposed a radical thermodynamic cycle for the new engine—a bypass ratio of 8:1, which offered a quantum improvement in propulsive efficiency. For such a high bypass ratio to work, the engine must have a compressor with a pressure ratio of 25:1, far higher than any yet in service. Such a compressor in turn requires a very high gas temperature into the turbine: 2300°F. If all that could be achieved, if all the parts would hang together and maintain their performance levels over the thousands of hours of transport operation, the new engine would have a specific fuel consumption an order of magnitude lower than anything then in service.

The Powerplant Laboratory at Wright–Patterson Air Force Base accepted the thermodynamic theory but doubted that GE could make it all happen. To prove its point, GE built a one-third scale technology demonstrator engine, the GE1/6. The engine ran and demonstrated a minimum specific fuel consumption of 0.336 lb fuel/lb thrust per hour, less than half that of the best engines then in service. Pratt & Whitney also built a demonstrator engine for the C-5A, with a bypass ratio of 3:1 and lower pressure ratio and gas temperatures than GE's. Boeing liked the approach proposed by GE, while Douglas and Lockheed were more comfortable with the conservative Pratt & Whitney cycle. This was GE's first real dual-rotor front fan engine, with variable stator vanes on the high-pressure compressor. The GE1/6 was tested some 14 years after Pratt & Whitney's first J57 prototype and 11 years after GE's first demonstration of the variable stator. It had taken all of that time for GE to overcome its own biases and marry the best of Pratt & Whitney's architecture to its own. From that time on, all of GE's large military and commercial engines were of the dual-rotor front fan configuration with variable stator vanes, as were Pratt & Whitney's. Rolls–Royce avoided variable stator vanes for a time by using three rotors in its turbofan engines, but ultimately also adopted variable stators.

GE's second gamble was to propose a total package price at a razor-slim profit margin, reasoning that winning at a modest profit would be much better than not winning at all. In 1965, with a clear performance superiority and a very attractive commercial proposal, GE won the competition and began development of the TF39 engine for the C-5A. The $459 million contract was the largest order GE had ever received for aircraft engines. The U.S. Air Force selected Lockheed to design the aircraft and directed it to use the GE engine. However, an unexpected complication arose in that Lockheed had quoted a package price for the C-5A that it was unable to achieve. Rather than have Lockheed go under and lose the entire aircraft program, the U.S. Air Force reduced the number of aircraft in the program from 120 to 81 for the same price and accepted a limit on the aircraft's maximum allowable load factor until Lockheed could strengthen the wing under a separate contract. Some years later, the U.S. Air Force bought a second batch of C-5 freighters to bring the total up to the originally stipulated 120, and GE put the TF39 back in production. In all, GE delivered 680 TF39 engines and, by rigorous attention to cost made a small profit. A few years before, GE had lost $80 million building 408 CJ805 engines.

GE designed the TF39 to military specifications, while keeping in mind the needs of the U.S. airlines with whom the TF39 design had also been reviewed. Because the TF39 was destined to be a transport engine, the life requirements for its parts were much longer than would have been the case for a fighter or bomber engine. Military and commercial customers use different systems of product support, and those are reflected in special design features for inspection and maintenance. But the mechanical design criteria for long component lives are the same whether the part goes into the engine for a military freighter or one for a commercial airliner.

Later, when GE began the design of the CF6-6 commercial engine, it started from the base line of the TF39's excellent gas generator or core engine, and made significant changes to its proven thermodynamic and mechanical design. Because airliners would cruise at higher speeds than the C-5A freighter, GE simplified the fan[4] and lowered the bypass ratio, raised the cycle's gas temperature higher than it was in the military engine, improved the life and maintainability of many of the parts, and applied new materials and better cooling technology to incorporate the experience of the TF39. The technical success of the TF39 set a performance benchmark for turbofan engines and gave GE credibility as a designer and manufacturer of commercial engines.

---

[4]The TF39 fan had a revolutionary stage of half-span in front of the main fan. The CF6 fan eliminated this half-span stage but added a "quarter" span stage behind the main fan as a booster to increase pressure to the inlet of the compressor.

Chapter 4

Back in the Commercial Engine Business

ALTHOUGH Boeing had lost the U.S. Air Force strategic freighter competition to Lockheed, the Boeing freighter became the design basis for a new large airliner, the 747, capable of carrying up to 400 passengers on transoceanic routes. With a fuselage far wider than that of existing airliners and using the new fuel-efficient high-bypass turbofan engines to cruise at a Mach number of 0.9 (in comparison, the cruise Mach number for the C-5A was 0.76), its direct operating cost per seat–mile would be substantially lower than that of the 707 or DC-8. Its productivity promised to be so great that it could carry more passengers across the Atlantic in one year than a large ocean liner such as the QE2. Pan American Airways saw the 747's potential at once and gave Boeing an order for 25 aircraft, locking up Boeing's first 25 delivery positions and giving PAA a lead of one year or more over other airlines.

Pratt & Whitney had proposed a conservative engine design with a bypass ratio of 3:1 for the C-5A, and had lost. Now Pratt & Whitney had looked at the features of the TF39 that had allowed GE to win the freighter engine competition and responded quickly to the Pan American specification by launching the design of the JT9D, a completely new engine with a bypass ratio of 5:1, a high pressure ratio compressor and an air-cooled turbine.

Boeing asked GE to propose an engine of the TF39 type for the 747. The TF39 had been optimized precisely to the C-5A, its fan designed with no consideration to low noise—not an important requirement for a military freighter but very important for an airliner—and a core too small to produce the thrust required to cruise on a hot day at the specified 0.9 Mach number. GE faced a dilemma: should it develop a new fan and a new core compressor, in effect a second new high bypass turbofan in parallel with the TF39? Gerhard Neumann was prepared to take big technical risks, but he was conservative in business matters. He believed that GE had too many programs under way to be able to develop a new commercial engine and still do justice to the TF39. A lot of money would be riding on the success and cost control of the TF39; finally, PAA and Boeing asked for a delivery date that Neumann believed GE could not achieve. Several of the young managers on his staff disagreed with him on both counts.

For a short time GE even considered joining forces with Rolls–Royce to beat Pratt & Whitney, but Neumann was not interested in working on the triple-rotor engine design proposed by the British. He and Parker expressed all these reservations to Boeing, but PAA would not accept a change of the delivery date. Reluctantly, GE decided not to propose an engine for the Boeing 747, and Pratt & Whitney's new JT9D became its baseline engine.

Circumstances change. By the end of 1970, it became apparent that the U.S. Congress would end government funding for SST development, and that would end work on the GE4 engine. Military procurement had also been reduced. The U.S. Air Force had already canceled the programs to develop the B-70; a Mach 3 bomber; and the F-108, a Mach 3 fighter (one interesting technical reason was that the aircraft wing glows dull red on the leading edge at Mach 3, making it a perfect target for heat-seeking missiles), and with them GE's development of the J93 engine. Most galling to GE, the U.S. Air Force selected Pratt & Whitney to develop the F100 for the new F-15 fighter, despite Pratt & Whitney's problems with the TF30 in the F-111, while GE received only a development contract for the F101 engine for the B-1 supersonic bomber (a contract that would turn out to provide the key to GE's long-term success). Although it seemed little more than a consolation prize at the time, winning the F101 turned out to be a prophetic event: the F101 core engine compressor served as the basis for a number of subsequent engines, military and commercial. At that point, Neumann conceded that GE now had the resources required to develop a new commercial engine based on TF39 technology. With Pratt & Whitney entrenched on the Boeing 747, and Boeing showing no great enthusiasm for certifying the aircraft with an alternate engine, the new GE engine would have to be aimed at other airliners. In fact, Douglas and Lockheed were studying such new aircraft and the market for them. While the 747 promised to revolutionize the economics of air transportation with a steep reduction in direct operating cost per seat–mile, it was too big for the passenger load on many of the intercity routes in the United States. When U.S. airlines were still regulated, fares were officially approved by the Civil Aeronautics Board to cover actual operating costs plus a profit. Airlines competed largely on the basis of flight frequencies and customer service on the important routes between major "city pairs"; whether an aircraft flew full or empty had only a secondary effect on profit. When deregulation came to the United States in 1968, airlines began to compete on ticket price also and changed their route structures radically to the system of "hubs and spokes," a structure that often required aircraft of different sizes. At the same time, the critical need to make reductions in aircraft–mile operating costs put emphasis on standardization of aircraft types within an airline fleet. Some of the big carriers in the United States expressed a need for a 250-passenger airliner with the same direct operating cost per seat–mile as that of the larger 747.

Frank Kolk, the chief engineer of American Airlines, asked for proposals for a big twin-engine airliner with a fuselage as wide as that of the 747. The airlines expected passengers to prefer the roomy cabins of the wide body, and the new high bypass turbofans would give much lower specific fuel consumption and, for their size, less noise. Douglas and Lockheed responded with aircraft designs, and Rolls–Royce offered a preliminary design for an engine, the RB.207. The aircraft kept getting bigger and heavier during the preliminary studies, as is the normal case. Because a twin-engine airliner must be capable of flying successfully on the power of one engine, it became apparent that Kolk's "wide body" would require an engine with a maximum thrust rating of about 50,000 lb.

GE began design of a commercial engine based on the TF39 core but, at 40,000 lb thrust, the CF6-6 was too small for the big twin. Douglas Aircraft, GE learned, was uncomfortable with the concept of the twinjet and believed that a bigger trijet airliner with longer range would be more versatile. The top managers of several large airlines agreed. The CF6 would be a good fit for the trijet. GE's preliminary design manager, Jim Krebs, made the rounds of the major airlines to persuade them that a three-engine airliner would be better suited to their needs than a twinjet. GE had a persuasive argument: the FAA's certification specification mandates that an airliner be able to take off safely after failure of one engine at a critical point during the takeoff run. The power available to the pilot in such an emergency on a twin-engine aircraft is 50% of the original takeoff maximum; on a three-engine aircraft, two-thirds of the maximum takeoff power remain available after the failure of one engine. Runway lengths at the major airports are fixed, literally cast in concrete. With a greater fraction of its maximum installed thrust available, an airliner equipped with three engines can continue to take off safely after the failure of one engine, *at a greater weight* than one equipped with only two engines. A greater takeoff weight allows more passenger seats in the trijet, an immediate economic advantage over the twin. Add the additional capacity for cargo available in the belly hold of a wide-bodied jet, and the advantage for the trijet widens. GE calculated that the trijet could generate as much as 6% additional revenue and the airlines agreed, setting the stage for the design of two wide-body trijets, the Douglas DC-10 and the Lockheed L-1011. GE had an engine of the right size, and Frank Kolk angrily accused GE of having killed his twinjet.

On the Boeing 707 and Douglas DC-8, the airlines had become used to working with Pratt & Whitney and Rolls–Royce. Hearing that the airlines were not always satisfied with Pratt & Whitney's engines or with its product support, GE went to American and United Airlines, the two biggest carriers in North America, and to KLM in Eu-

---

[5]Nevertheless, as bigger engines later became available and demonstrated their reliability, airlines chose twin-engine airliners such as the Boeing 767 and Airbus Industrie A330 over trijets because they required less capital investment.

rope, and asked them to give GE a chance to compete for the engine in the DC-10. Douglas had impressed on GE the importance of sharing the risk of potential penalties for late delivery or any performance deficiency; GE was confident of its engine design and promised to make a long-term commitment to first-class product support. GE's corporate financial strength and reputation for standing behind its commitments on the CJ805 helped to validate the guarantees by Douglas and GE Aircraft Engines. The airlines, preferring not to be captive to any one engine company, also liked the idea of another supplier. Naturally, GE would have to go through a rigorous technical evaluation, but the airlines could see some benefit from an engine competition and from better product support.

At the time, Pratt & Whitney may not have had much faith in the concept of the trijet, for which its JT9D was not a good fit. Pratt was deeply involved in the Boeing 747 program, and there the JT9D engine was facing very serious problems. Part of the problem was that the airframe of the 747 weighed more than predicted, and needed more thrust to meet the PAA specification. When Pratt & Whitney raised thrust to the level required, turbine blades began to fail and the casings of the engine became distorted ("ovalized") at full power, allowing hot gas to leak around the turbines; this raised fuel consumption and reduced available thrust. Pan American refused to accept the aircraft unless they met the specification. At one point, more than 30 complete 747 airliners sat at the ramp at Boeing without engines, huge blocks of concrete hanging from their wings to substitute for the static load of the missing engines. Pratt & Whitney was paying penalties to Boeing and Pan American while it scrambled to correct the problem and was in no position to tackle another application. But Rolls–Royce, whose RB.207 had not found an application, was still a competitor for the trijet engine. Rolls–Royce proposed a somewhat smaller version of the RB.207, the RB.211. Lockheed favored this engine for its Tristar.

Neumann and his engineers worked closely with Douglas to optimize the performance of the CF6 installation in the DC-10 (see Fig. 5), spending as much time reassuring Douglas of GE's dependability as on the engine's technical merits. Validation came from an unexpected quarter when Douglas was taken over by McDonnell Aircraft

Fig. 5 Brian H. Rowe with a CF6 Engine ready to be trucked from Cincinnati to Douglas in California. (GE negative)

Company, the builder of the very successful F-4 Phantom fighter, which used two GE J79 engines. GE and McDonnell Aircraft Company had learned to work well together as an effective team.

Changing the number of engines for the new 250-seater from two to three may have suited the thrust of the GE CF6, but it also created a challenging problem for the installation of the center engine. Two of the trijet's engines are mounted conventionally in pods, one under each wing, while the third sits on the airplane centerline in the rear. Mounting the third engine behind an S-shaped duct in the tail of the fuselage, as Boeing had done in the 727 and as Lockheed was doing in the L-1011, would cost Douglas two complete seat rows against the Lockheed airliner because the two-spool CF6 was longer than the three-spool RB.211. With GE's installation engineering support, Douglas decided to mount the third engine in the vertical stabilizer of the aircraft tail, eliminating the S-shaped duct and its aerodynamic friction losses. That increased structure weight a little but also saved the two seat rows.

As the DC-10's weight grew, thrust had to be increased. The original concept of a TF39 derated for commercial service would no longer suffice. To get more thrust, GE made the fan diameter smaller and the fan pressure ratio greater; to meet the new noise specifications, it eliminated the inlet guide vanes. GE designed a completely new fan with a nonconstant energy distribution from hub to tip, solid titanium blades (Rolls–Royce used graphite/epoxy blades), and a booster stage behind the fan. The fan's acoustic design included substantial clearance between the blades and the outlet guide vanes and a plastic composite honeycomb shroud incorporating sound-deadening devices called Helmholtz cavity resonators, which work on the same principle as a pistol silencer. The core of the engine was still the same as that of the TF39, and the CF6 was able to use five stages of the TF39's six-stage low-pressure turbine. Even though it was longer than the Rolls–Royce RB.211, the CF6 weighed less and had a lower specific fuel consumption by virtue of its higher pressure ratio and turbine inlet temperature.

The two aircraft designs competed head-on, Lockheed L-1011 against DC-10. Lockheed needed a minimum number of firm orders to launch airplane development and production. Help came from an unexpected quarter. In 1967, a hitherto unknown British aircraft broker, Air Holdings, Ltd., ordered 50 L-1011s from Lockheed and specified the Rolls–Royce RB.211 engine. The British government offered export credit guarantees. At one stroke, Lockheed had its minimum launch number and was able to go ahead. It was then revealed that Rolls–Royce owned 50% of Air Holdings, Ltd., which quickly sold its contract for the 50 aircraft to TWA and Eastern Airlines, then two of the U.S. major airlines. The loss shocked Douglas and GE and left the future of the DC-10 and of GE's CF6 engine in the hands of American and United Airlines, the next major airlines to consider trijets. After a furious sales campaign, both airlines selected the DC-10 with GE's engine over the L-1011 with Rolls–Royce. Had the DC-10 lost again, there would have been serious doubt whether Douglas and GE would continue with it and the CF6 engine, and the trijet market would have belonged to Lockheed. That possibility may have been a factor in the two airlines' decisions. Both had used Douglas airliners in the past and wanted to be able to rely on more than one supplier for large airliners. Douglas was a major producer of airliners at the time. Although Lockheed's airliner experience had begun with the Vega and Electra before World War II and continued with the Constellation and L-188 Electra turboprop after the war, the L-188 had suffered from a serious safety problem with its engine nacelle. Douglas must have been seen as posing a lower risk. GE's reputation of standing behind its CJ805 engine through thick and thin helped persuade American Airlines, and GE's financial strength was no doubt an important factor in persuading United Airlines.

These were crucially important orders. They set an example for other airlines, some of them with less capability than American or United Airlines to analyze the technology and economics of a new class of airliners. Soon after, Delta, Continental, and National Airlines also ordered the DC-10 with CF6 engines.

Rolls–Royce had already run into serious trouble in the development of the RB.211. The original graphite–epoxy composite fan blades could not withstand hailstones and bird impact and had to be replaced with titanium blades. This led to a complete redesign of the entire fan disk and by necessity of the fan shaft as well. The RB.211 is a triple-rotor engine, very short and stiff, with stable aerodynamic performance, but inherently mechanically complicated. The front fan is driven by its own turbine, independent of the two turbines driving the low, and high, pressure compressors. That allows the use of fewer booster stages behind the fan and fewer stages in the low-pressure turbine. The RB.211 proved to be heavier than its dual-rotor competitors, and the new titanium fan increased its weight even more. At the end of 1970, admitting that it could not meet the weight guarantees for its engine and that it had incurred huge development cost overruns, Rolls-Royce was forced into financial receivership. Because of the firm's prominence and importance, it was taken over by the British government, which repudiated the performance and price commitments Rolls–Royce had made to

Lockheed and the airlines, and the penalties imposed on Rolls–Royce by Lockheed in case of late delivery. The British government in effect told TWA and Eastern Airlines that they had to pay whatever it cost to get RB.211 engines or do without. Grudgingly, the airlines agreed to pay the price.

Lockheed itself came under powerful financial pressure from TWA and Eastern Airlines and was rescued only by a loan guarantee from the U.S. government. GE testified to the U.S. Congress at the time that it could see no justification for a U.S. loan guarantee to rescue Lockheed with a British engine when an American engine was available without government financial aid. Lockheed and the airlines were not pleased; the British government, infuriated, for years afterward looked on GE as an enemy. Some 30 years after this high-stakes international poker game, American and United Airlines survive as major carriers in America, Eastern Airlines has disappeared, and TWA remains in serious financial difficulties. The L-1011 became a successful aircraft, but Lockheed no longer makes commercial airliners. It remains a successful manufacturer of military aircraft and aerospace electronics, having merged with the fighter division of General Dynamics, with Martin–Marietta, and with GE/RCA's aerospace group. Lockheed's attempts to diversify into civil airliners with the L-188 Electra and the L-1011 TriStar must be judged expensive commercial failures.

Product superiority is always a key to market advantage in commercial aviation. To break away from the L-1011, Douglas decided to design a larger version of the DC-10, the intercontinental-range DC-10-30, challenging the larger Boeing 747 with a very economical trijet. As the design of this aircraft evolved, the thrust required from the engines grew to 48,000–50,000 lb. GE's advanced product planning operation had previously studied concepts to increase CF6 engine thrust to 45,000 lb by adding two or three booster stages in the core flow path behind the fan to increase core compressor mass flow and pressure ratio. Pratt & Whitney used booster stages on its turbofans but encountered booster stall during transient power changes because of mismatched mass flow between the booster and the low-pressure compressor. GE's compressor designers believed that they could avoid such problems by proper scheduling of the variable stator vanes, a technology GE had pioneered, and by using bleed doors aft of the booster stages to dump excess airflow into the bypass stream. But what the new and bigger DC-10 required exceeded the realistic capabilities of such possible fine tuning. A more powerful CF6 required major changes in pressure ratio, compressor discharge temperature, and turbine inlet temperature. Arthur Adamson, the manager of CF6 engineering, came up with the technical solution: remove the last two stages of the low-pressure compressor, he recommended, and increase the mass flow and pressure ratio of the booster. Many changes would have to be made to the flow functions aft of the compressor at the cost of a major development program. For example, the entire turbine section would have to be enlarged. But GE would be able to keep the fan diameter to 86.4 in., the same as that of the CF6-6, preserving the external diameter of the engine. The design could also stay within reasonable limits of compressor discharge and turbine inlet temperature. The resulting CF6-50 (for 50,000 lb of takeoff thrust) had a bypass ratio of 5:1, an overall pressure ratio of 31:1, and turbine inlet temperature of about 2350°F. It also met the thrust requirements of the new DC-10-30.

Before GE committed certification of the CF6-50 to Douglas, Neumann formed a task force to consider alternative approaches to achieving greater thrust. Some of his engineers believed that scaling up the core of the CF6-6 for greater mass flow, and perhaps scaling up the fan, would be a surer strategy and permit more margin for thrust growth in the future. Such an approach would also take more time and a more expensive development program. Adamson and the advanced product planning engineers prevailed and, to make sure that they were on the right track, modified the CF6-6 fan test rig with booster stages to get quick test results to validate their aerodynamic predictions. GE decided to go ahead with Adamson's proposal and produced the CF6-50, the first of the big engines in the 50,000-lb thrust class. It entered service with a rating of 48,000 lb and grew to 50,000 and then 53,000 lb as component efficiencies were improved. It had the best thrust-to-weight ratio and the best installed performance of all the competing engines because of the combination of the 86.4-in. diameter fan with higher overall pressure ratio and higher turbine temperature than its competitors.

European airlines, generally smaller than the big U.S. carriers such as American or United Airlines, have organized themselves into syndicates to select common equipment. By pooling their purchases they hope to get a better price; by rationalizing maintenance among themselves, they hope to improve their productivity and reduce their costs. The first such syndicate, called KSSU, was composed of KLM, Swissair, Scandinavian Airlines, and UTA. Many of them had already ordered Boeing 747s but also wanted smaller aircraft for their "thinner" long-range routes with transoceanic segments. They had some interest in a trijet, and Douglas proposed the DC-10-30. GE's Burgess was sent to Europe to support Douglas in marketing it. He continued to cultivate personal friendships started over the CJ805 with senior managers at Swissair, KLM, and SAS. Over time, as had been predicted in 1953, his friends had become chief executives of their airlines. Because an airline engine model is in service for 20 or 30 years, taking several years to develop personal relationships is not out of proportion. GE enjoyed a high degree of technical credibility with KSSU, and Burgess now had the personal entree to its top executives. KSSU picked the DC-10-30 with the CF6-50. The sale was soon followed by one to the ATLAS syndicate: Air France, Alitalia, Lufthansa, Sabena, and Iberia. The DC-10-30 and its CF6-50 engine were off to a strong start in Europe.

But success was not always assured. Northwest Airlines also needed such an aircraft. Neumann himself gave Northwest Airlines' management an enthusiastic sales presentation, during which the airline chairman appeared to doze. He opened his eyes at the end and said, "Why don't you stick to light bulbs? GE is good at making light bulbs. Leave aircraft engines to Pratt & Whitney." Shades of the turbosuperchargers! Northwest Airlines did select the Douglas DC-10, but with a version of the Pratt & Whitney JT9D, and Pratt & Whitney agreed to pay for the cost of aircraft certification. It turned out to be a bad decision. No other airline ordered that version of the DC-10. The Pratt & Whitney engine had more than its share of troubles, and there was only a small population of such engines to bear the cost of fixing them—the exact economic problem GE had faced with the CJ805.

GE's dogged product support had been the major redeeming feature of its first commercial engine, the CJ805, which had a reputation for mechanical fragility. To take care of its customers, GE had devoted a lot of effort to product support, with generally good reception from the airlines. But the CJ805 was swamped in the market by Pratt & Whitney's JT3 engine, the powerplant for the popular Boeing 707 and Douglas DC-8 airliners. GE lost some $80 million on the CJ805 program, a result that did not delight its corporate management. On the new CF6 engine, GE was determined not to repeat the mistakes that had given the CJ805 a bad reputation inside the company. Anyone who had been active in the CJ805 program was still tainted by that stigma; so outsiders were brought in to run all phases of the CF6 program. Some of them came without any airline experience and, when they prepared the business plan for the CF6, they failed to include a budget for product support. GE quickly pulled its experienced people back into the CF6 program.

Richard B. Smith, who had been product support manager for the CJ805 and later for the business jet engines, was brought in to set up a marketing organization for the CF6. GE wanted a systematic way to train its salesmen, back them up with sales engineering and financial analysis, draft and negotiate contracts and warranties, and provide sales financing. GE already had operations and service engineers working with the airlines and had learned what was needed for marketing and product support with the CJ805 and the business jets. The challenge now was to get capable people into the organization and train them. GE gave the responsibility for organizing CF6 product support to Walter Van Duyne, who had earned his spurs with Wright Aeronautical Corporation and had been hired by GE for the CJ805. Van Duyne laid out an organization: so many people for repair engineering, spare parts, warranty administration, and field service. He specified a training program, lasting up to one year and six months for service engineers who had never been in the field before. And he asked for the best people in GE's military field service engineering operation. Brian Rowe, the CF6 project manager, blinked but gave him the budget and his blessing to go ahead, convinced that meeting promises and treating customers fairly would give GE a reputation that would help the business grow. A key to this successful approach was direct contact with top management. Rowe and Neumann told each of the field service people, "You work directly for us. If you have a problem getting support from the factory, you call me direct!" Which they did. One of the basic differences between Wright Aeronautical Corporation and GE was that Wright Aeronautical Corporation's top management harangued the people in the field and constantly reduced their operating budget, while GE's management at the top regarded the people in the field as a vital resource for maintaining contact with the customers and keeping them satisfied. The customer service people had management support and knew that GE had the resources to keep its promises—even when the commitments were, on some occasions, ill-considered. Credibility was the important objective.

In the beginning of CF6 airline service, Van Duyne's field engineers met every flight until the engine had proved itself in operation. Every morning at 7:30 a.m., Rowe met with his staff and Van Duyne would report service problems that had arisen during the previous day. Rowe and his managers made assignments for corrective actions on the spot. Weekly, Van Duyne met with the manufacturing vice president to review critical spare parts needs. Manufacturing usually puts greater priority on getting complete engines delivered than on making spare parts, but the vice president considered himself part of the GE team that had promised the airlines to keep their planes flying. GE's product support people felt themselves charged with the responsibility for customer satisfaction. They were convinced that GE had given them the authority and resources to build a working partnership with the customer. Sometimes it was not clear for which side they were working: when GE occasionally raised spare parts prices, the product support people complained along with the airlines. The overall effect was exactly what GE had wanted: the airlines were satisfied that all of GE, from the field service engineer to top management, stood ready to do what GE had promised.

CHAPTER 5

Airbus Industrie and the Birth of Revenue-Sharing Strategy

EUROPEAN governments, like those of many countries, like to foster the growth of their own technology industries. These, the governments believe, add high value to the manufacturing process, create products for export, and may also reduce the need for imports. In the late 1960s, their attention focused on airliners, which were being ordered mostly from Boeing, Douglas, or Lockheed. There were sensible reasons for this: at the outbreak of World War II, the Douglas DC-3 was arguably the most advanced and economically efficient airliner in the world. Thousands were built during the war for the U.S. armed forces, and U.S. manufacturers continued development of larger airliners with longer range, such as the DC-4 and Lockheed Constellation, even while they were devoting most of their resources to the production of warplanes. At the end of the war, the United States had the only undamaged national economy, and its airline networks grew much faster than any other nation's. The U.S. aircraft manufacturers were able to develop high efficiency and low costs in satisfying the national market for airliners, and they kept developing larger and more productive aircraft, whose economic performance and reliability set the standard for the world outside the former USSR. These aircraft were equally attractive to airlines in other countries, allowing U.S. manufacturers to build up a seemingly unassailable competitive lead over any of their competitors. The governments of France and Germany decided to cooperate in supporting the development of European airliners, breaking the grip of this U.S. monopoly. In 1968, government planners led by France and closely supported by Germany began studies of a new European airliner, studies that in 1970 resulted in the formation of the Airbus Industrie syndicate to design and build airliners specifically for the short- and medium-range city pairs of Europe (ultimately, Airbus Industrie expanded its product range to include large long-range airliners as well). Most European airlines were owned by their governments and formed a captive market. The British hesitated about joining Airbus, just as they hesitated at that time about joining the European Economic Community, preferring instead to develop their own airliners for British and world markets. (Some years later, British Aerospace joined Airbus Industrie as a private partner.)

The first Airbus Industrie design was the A300, a wide-body twin somewhat smaller in seat capacity than the Lockheed L-1011 and Douglas DC-10, looking in fact a lot like Frank Kolk's twinjet. There was no equivalent American aircraft. During meetings with Professor Hoeltje, the technical director of Lufthansa, and Henri Ziegler, the president of Airbus Industrie, Gerhard Neumann was told that the A300 faced a political mandate. To the greatest extent possible, every nut and bolt on it was to be European. With a European engine, the Rolls–Royce RB.207, available and the German government generally favoring an association with Rolls–Royce, there was no room for an American engine, even though the French government and industry leaned toward Pratt & Whitney, which at that time owned 11% of SNECMA's shares. The A300 would satisfy a specific market in Europe and would compete for similar airline routes in the rest of the world, even in the United States.

GE had no great expectations for the A300—a new market entrant such as Airbus with no established reputation might sell perhaps 100 aircraft. But GE had made a major commitment to the commercial market with its CF6 engine for the DC-10, and European airlines had ordered the DC-10. Getting the CF6 into the A300 could support more DC-10 sales to the same airlines, and even a small A300 program would offer GE some gain at acceptable incremental cost. Gerhard Neumann set out to gain the confidence of Ziegler at Aerospatiale, the French Airbus Industrie partner, and of Dr. Weinhardt at Deutsche Airbus in Germany. Airbus Industrie was more than aware of its status as a newcomer against Boeing and Douglas and hypersensitive to the possibility that an American engine supplier would favor American airframe manufacturers over the Europeans if their airplanes were competing for the same sales opportunity. Neumann needed to convince Airbus Industrie of GE's genuine commitment to the A300 and impartial responsiveness to Airbus Industrie's marketing needs. Because Airbus Industrie had the only twin, against the Douglas trijet and the Boeing 747, this was not an impossible task. The engine nacelle was one competitive factor. Rolls–Royce offered a complete nacelle integrated with its engine with the argument that it is difficult to separate the contributions of engine and nacelle to overall aircraft performance and noise levels. Neumann suggested to Airbus that the successful DC-10 nacelle with the CF6-50 engine might fit the A300 as well and save Airbus Industrie a lot of experimentation. Neumann promised to discuss the subject with the president of McDonnell-Douglas.

Neil Burgess, in Miami with a Douglas team marketing the DC-10 to National Airlines, told Jackson McGowen, the Douglas marketing manager, that working with Airbus Industrie could strengthen the market for the DC-10 in Europe and enhance the Douglas reputation with the governments of France, Germany, and Spain, which owned Airbus Industrie. McGowan called Donald Douglas Jr. and got his approval to offer the nacelle data package to Airbus Industrie. Although the A300 could have used the GE CF6-50 engine with the complete Douglas pod and mounting strut, Airbus Industrie decided to design its own strut and pylon and agreed to pay Douglas $7.2 million for the rights to use the nacelle drawings. GE advanced the money to Airbus Industrie, to be repaid at the rate of $60,000 for each engine bought until the money had been paid back with interest. GE also persuaded Rohr, its supplier of the CF6 thrust reverser for the DC-10, to work with Airbus Industrie. Like GE, Rohr saw an incremental market and agreed to locate a thrust reverser assembly shop in Toulouse, France, next to the Airbus Industrie assembly building.

Although all of this served to strengthen GE's relationship with Airbus Industrie, it did not satisfy the industrial policy objective of the French and German governments for a "European" engine for the A300 because it created no work for the European engine industry, in 1968 largely government-owned, as were the airframe manufacturers and the airlines. GE decided to enlist the biggest engine manufacturers as strategic partners: SNECMA in France and Motoren- und Turbinen-Union (MTU) in Germany, the chosen instruments of their governments for large aircraft engine programs.

Neumann's reputation was well established with the European engine manufacturers. He had been deeply involved in European licensed production of the GE J79 engine for the F-104G Starfighter, for which more than 2000 engines were built in Europe and, subsequently, another 475 for the German F-4G Phantom. The J79 program helped reestablish engine manufacturers after World War II in Germany, Belgium, and Italy with the then-latest American technology. The license fees paid by the European governments to GE had been quite reasonable, lower than they had anticipated; at the same time GE was able to fill substantial orders for complete engines, subassemblies, and parts to protect European schedules until their own manufacturers were in full production. The program had been profitable for all of them and had created personal relationships of trust in technical collaboration between GE and European engine companies.

Rolls–Royce with its proposed RB.207 was GE's competition for the A300. If the Rolls–Royce engine were selected, SNECMA and MTU might be asked to participate, but only as subordinates to Rolls–Royce, in much the same way as the French airframe company Aerospatiale dominated Airbus Industrie. GE decided to offer SCEMA and MTU an association of near-equals. The big problems were productivity and comparative costs. It was extremely difficult to make and sell a CF6 engine part in Europe at a cost significantly lower than that of GE. Greater American productivity and volumes, and higher social costs in Europe, combined to make the American industry very competitive in cost. Price, on the other hand, is whatever the seller chooses to ask.

Edward Hood, the general manager of GE's commercial engine division, invented a novel concept to solve the problem of different manufacturing costs. He proposed that SNECMA and MTU make

40% of the parts of the CF6-50 engines ordered for all A300 aircraft no matter where the aircraft were sold, not only for those bought by French and German airlines. GE would pay *not a subcontract price* (which would normally have to be lower than GE's own manufacturing cost) *but a proportional share of the total engine price*, the so-called revenue share. If SNECMA and MTU were efficient, their manufacturing costs could reasonably be lower than this revenue share. European manufacturers would be paid their portion of the engine selling price in U.S. dollars, the normal currency for airliner sales, while their manufacturing costs were incurred in their own currencies. When later the dollar-to-franc exchange rate went strongly against SNECMA, GE would agree that SNECMA should make more parts of fewer different part numbers for other CF6-50 engine applications as well as the A300, for example for DC-10 and 747 aircraft. That increase in volume allowed SNECMA to bring down its unit manufacturing costs. The principle throughout was that this was a partnership to benefit all the participants: no partner would drive the other into the ground if external circumstances changed. GE also offered MTU a patent license for its unique process for drilling the deep holes needed for cooling turbine parts, which MTU could use for other programs of its own. In return, MTU agreed that it would not join forces with any other engine manufacturer for the A300.

Rolls–Royce and Pratt & Whitney were of course also competing for the Airbus Industrie engines. Pratt & Whitney had its hands full with the engine problems on the 747 and had no great expectations for the A300, perhaps because of its earlier experience on another French airliner, the Dassault Mercure, which used two JT8D engines, but of which only 14 were ever built. For the A300, Pratt & Whitney offered SNECMA manufacture of the JT9D under license, or conventional subcontracts if Pratt & Whitney made the engines; oddly, Pratt & Whitney did not initially agree to make free engines available to Airbus Industrie for flight tests, even though at that time Pratt & Whitney owned about 11% of the shares of SNECMA, having received them in return for licenses on JT8D commercial and TF30 military engine technology. Pratt & Whitney's equity position did not affect SNECMA's decision. In fact in later years, when

SNECMA increased its capital, Pratt & Whitney did not want to subscribe additional capital to a company that had by then become a major partner of GE, and saw its equity share in SNECMA diluted and reduced.

Rolls–Royce had concentrated on the RB.211 engine for the Lockheed L-1011 and, in 1969, was already wrestling with technical problems and cost overruns that would send it into receivership in 1971. The bigger RB.207 offered for the A300 existed only on paper. Although the Rolls–Royce Avon had been used in the Aerospatiale Caravelle, Rolls–Royce was focused on its prime program with Lockheed and all the attendant problems and could see little advantage in dissipating its resources on an unproved European airframe syndicate. Like Pratt & Whitney, Rolls–Royce refused at that time to promise free engines to Airbus Industrie for flight test. Ziegler called Neumann, who at once offered eight "free" engines for A300 flight test and certification, for which Airbus agreed to pay after it had sold 200 aircraft. To forestall a possible alliance between Pratt & Whitney and Rolls–Royce, GE even offered cooperation to Rolls–Royce on the Airbus Industrie engine. Rolls–Royce had ambitions to launch its own big engine later—ultimately the RB.211-524 for the Boeing 747—and decided not to have anything to do with the CF6-50: Rolls–Royce considered the GE revenue-sharing concept little more than a cunning plot to destroy SNECMA and MTU.

The initiative for a European airliner had come first from the French government. The French economy would have a big stake in the success of the Airbus Industrie program and in SNECMA once the A300 had been launched. GE's offer to SNECMA was not a make-or-break matter for the European airlines that had ordered the DC-10-30 and were now the natural market for the A300, but it was certainly very important to the French government. Could SNECMA succeed, could SNECMA make a profit, under the conditions proposed by GE? There were a lot of skeptics in Paris. While the government's technocrats may have reserved judgment, Airbus Industrie's Ziegler was impressed by Neumann's sincerity, by the performance of the GE engine, and by GE's successful development of the TF39 for the C-5A. Airbus Industrie picked the GE CF6-50 for the A300. SNECMA and MTU informed their governments that they supported this choice and planned to be partners in the manufacture of the engine.

The two governments gave their consent and Airbus Industrie received launch orders for the A300 from Air France and Lufthansa, quickly followed by orders from other airlines. The A300 was an efficient airliner. It became a commercial success because it filled an unserved niche in the market and because financing by the governments owning Airbus Industrie allowed the company to manufacture A300 airliners on a steady basis in advance of orders. Thus, Airbus Industrie could fill orders very quickly out of an inventory of A300 "white tails," completed airliners with no airline insignia painted on their vertical fins. SNECMA and MTU went on to make their shares of CF6-50 hardware in one of the most profitable programs they ever enjoyed, and GE became the dominant supplier of engines for aircraft designed by Airbus Industrie. Fifteen years later, Airbus Industrie overtook Douglas in global market share, and three years after that, at the end of 1996, the Boeing Company announced plans to take over Douglas. (The demise of Douglas is described in Chapter 22.)

Chapter 6

Engines for the Boeing 747

THE Boeing 747, with its spacious, comfortable passenger cabin and its high productivity, had quickly established itself as the premier aircraft type for high-density long-range routes and was used by all of the major American, European, and Asian airlines. Boeing had launched the 747 with the Pratt & Whitney JT9D and had not certified it with any other engine. Even though it had suffered through the problems of aircraft performance deficiencies caused by JT9D problems, whenever an airline asked whether it could get a 747 with a Rolls–Royce or a GE engine, Boeing always replied that another certification would be quite expensive.

The U.S. government wanted to replace "Air Force One," the president's executive Boeing 707, and to use several aircraft of the same type as airborne command posts. The obvious choice was the 747, modified with substantially more electrical power than the conventional airline model for its communications gear. The U.S. Air Force designated this new aircraft the E-4 and called for development of engines capable of generating high levels of auxiliary electrical power, a development program that would necessarily cover most of the technical work required for certification of the 747 with a new engine. GE decided to offer the CF6-50 with special gearboxes capable of generating the electrical power required. The U.S. Air Force had excellent results with the GE TF39 engine in the C-5A Starlifter, and welcomed the GE proposal.

The E-4 is a showcase application, and it was vital for Boeing (and of course for the White House) that the aircraft and its engines perform flawlessly. Thornton Wilson, known throughout the industry as "T" Wilson, had just become chairman of Boeing and called Parker and Neumann to ask them how convinced GE really was about the reliability of the CF6-50. They told him about the engine's excellent experience in the DC-10-30 and promised to provide the same level of support for the E-4's engines as GE had been providing for some years for its engines in the helicopter squadron that shuttles the president and his cabinet around Washington, for which GE had developed special product support systems. The U.S. Air Force selected the CF6-50. Its confidence in the product was vindicated when the E-4 went into service. "T" Wilson himself was impressed enough to suggest to Parker that Boeing would pay for half the cost of civil certification of the CF6-50 on the 747 if GE paid the other half; then Boeing could offer the airliner with GE engines whenever the right opportunity presented itself. Parker agreed, even though Boeing already had orders for more than 200 747s with Pratt & Whitney engines and might sell no more than 25 with the GE CF6.

The first opportunity came quickly enough. Lufthansa was already using the Douglas DC-10-30 and Airbus A300, both with CF6-50 engines. Putting the same engine into its Boeing 747s made clear economic sense: GE claimed an advantage in fuel consumption over the Pratt & Whitney JT9D. But there would also be savings from having only one type of engine in the airline inventory and benefits in reliability from the familiarity of pilots and mechanics with a single model. Neumann and his marketing people worked hard to point out the advantages and convince Lufthansa that GE was very serious about supporting its customers. There was no need for commercial puffery. Lufthansa was using CF6-50 engines in part of its fleet and Pratt & Whitney JT9D engines in its 747s; the comparative data on performance, operating reliability, and maintenance costs were right there in the airline's own records.

In a momentous commercial decision, Lufthansa replaced its older 747s with new ones using the GE engine. That had wide impact in the industry because Lufthansa had a reputation for thorough engineering and meticulous analysis of operating costs. Soon Air France and the other KSSU and ATLAS airlines followed suit for the same reasons. So did 14 smaller airlines that took their technical lead from the bigger carriers, for example Finnair and Air Afrique. Later, "T" Wilson joked that he should have asked GE for $100,000 for each 747 sold with GE engines above the 25 Parker had anticipated. In total, 280 of the 1,290 747s ordered had GE engines, 150 Rolls–Royce engines, and 860 engines from Pratt & Whitney. GE had now succeeded in getting its CF6 engine into the twin, the trijet, and the four-engine airliners used by most of the major airlines.

GE's competitors did not fade away, of course. Other airlines, such as Singapore Airlines, continued to buy Pratt & Whitney and Rolls–Royce engines to take advantage of the fierce price competition between the engine manufacturers. Having added the CF6, Boeing also certified the 747 with the Rolls–Royce RB.211-524, which was then ordered by British Airways, Cathay Pacific, Qantas, and Air India. Some airlines, such as Northwest and Japan Airlines, continued to use Pratt & Whitney engines because of their years of experience with that manufacturer. Boeing concluded that its market was enlarged by the fact that it was able to offer the 747 with any one of three engines a customer might prefer. Airbus Industrie noted this phenomenon, and also began to offer its airliners with a choice of engines, even though the majority continued to be ordered with those from GE.

CHAPTER 7

The Big Twins Reshape Competition

SINCE the end of World War II, American airframe manufacturers dominated the international market for airliners. Of the leaders in the 1960s, Convair stopped making commercial aircraft after the 880/990 fiasco and Lockheed did the same after the Rolls–Royce technical and financial problems of 1970 threatened the viability first of the L-1011 program and then of the entire company. Boeing took a big lead with the 707 and 747, the latter in a market segment in which it had no competition; Boeing subsequently broadened its market position with the 727 for short- and medium-range routes. Douglas was in second place with the DC-8, the DC-9 and its derivatives, and the DC-10 trijet. To the Europeans, domination by the American industry was intolerable, whether by four firms or two. European companies have a long history of aircraft and engine design and consider themselves quite capable of coming up with competitive products. Industrial policy animated the French, British, and German governments to support their companies, and national pride was an important factor, as it is in the United States. In the late 1960s, preliminary design studies by Hawker–Siddeley and British Aerospace in England, SNIAS (later Aerospatiale) in France, and a syndicate of German airframe companies led to the concept of the A300 "Airbus" and the formation of Airbus Industrie to manage its design, certification, production, and marketing. The Dassault "Mercure," of which 14 were built, was also the outgrowth of such national planning, as were the SNIAS "Galion" and the British Aerospace BAC 3-11, both later dropped in favor of the European (rather than purely national) Airbus Industrie A300B.

The A300B found an immediate niche in the airliner market: it offered excellent performance and advanced control systems, a wide body preferred by passengers over the narrow-body 727 and DC-9, modern high-bypass turbofan engines from GE, capacity closely suited to medium- and short-range city pairs, and no doubt a very competitive price from Airbus Industrie, which was fighting to establish itself in the market. Eastern Airlines got preferred status from Airbus Industrie as the first A300B customer in the United States. Later, PAA was able to buy and lease A310 airliners on a similar basis. For the first time in many years, the U.S. industry had effective competition—particularly in Europe, but also in Asia and even in its own domestic market (as it turned out, neither Eastern Airlines nor PAA survived the rough-and-tumble competition in the United States during the 1980s and 1990s following airline deregulation).

Competition from Airbus Industrie put pressure on the price margins in every airline competition; this was especially damaging to Douglas, which could offer only the DC-10 trijet and the MD-80 twin derived from the DC-9, whose combined sales did not generate enough income to justify the development of a new Douglas airliner. Douglas found it difficult to match the performance of the newer aircraft types offered by Boeing and Airbus Industrie, and had to compete largely on price and on the basis of brand loyalty. Long-term Douglas customers such as Alitalia, Swissair, and Scandinavian Airlines System wanted to keep alive as much competition for airframes as possible, particularly if the costs of competition were borne by the competing manufacturers. As the chairman of Singapore Airlines said at the time, "I am responsible for the profitability of my airline—not that of my suppliers."

Douglas scrambled to create the semblance of a modern product line. The MD-80 was a lengthened DC-9 with refanned JT8D engines, an efficient derivative of the DC-9 with proven subsystems, great appeal to existing DC-9 operators, and manufacturing costs well down the learning curve. In the mid-1980s, in a joint venture with the Chinese Ministry of Aerospace Industry, Douglas put a version of the MD-80 in production in Shanghai in return for an order for 26 aircraft. But the Civil Air Administration of China (CAAC), the Chinese national airline, preferred to use Boeing airliners and grudgingly put the Douglas airliners into service on subsidiary routes. Douglas also modified the DC-10 with new engines from Pratt & Whitney or GE and added winglets, small canted extensions at the tip of the wing, to improve cruise lift. The new version was called the MD-11. The airplane was offered for long-range routes that did not carry enough traffic to justify the passenger-carrying capacity of a 747. Several loyal Douglas customers ordered the MD-11, but when Singapore Airlines refused to accept the ones it had ordered because they did not meet their guaranteed performance for range and fuel consumption, its market success elsewhere was limited. It is never easy to pinpoint with precision what causes such problems, but there was no doubt that the MD-11 had more drag than expected and some question whether the Pratt & Whitney and GE engines were as efficient as had been promised. The engine companies and Douglas pitched in to improve the situation and succeeded to a considerable extent, but the reputation of the MD-11 never recovered. Douglas needed a new aircraft design. There was talk of collaboration with Airbus Industrie, with the Europeans designing a new and more efficient wing for the MD-11 fuselage, but nothing came of it, perhaps because Douglas could not bring itself to accept such collaboration or perhaps because Airbus thought it had one of its two competitors on the run.

Douglas laid out the design of a new trijet, the MD-12, with competitive performance. There was only one problem: its development would cost more than $2 billion, money that McDonnell–Douglas was not able or prepared to invest on its own. John McDonnell, Mr. Mac's son and now the chairman of McDonnell, had previously stepped in to rescue Douglas Aircraft, but he was known to be skeptical about the profitability of an entirely new airliner. Although military aircraft production by McDonnell Aircraft was generating substantial cash flows, the company was facing several serious problems on two major military development programs—the Douglas C-17 freighter, intended as a replacement for the Lockheed C-141 Starlifter, and the revolutionary "stealthy" McDonnell A-12, intended as the replacement for the Navy's A-6 attack bomber. The U.S. DOD was exerting great pressure on Douglas about cost overruns in C-17 development, and the U.S. Navy was threatening to cancel A-12 development at McDonnell Aircraft because of rapidly escalating development costs. In fact, when McDonnell Aircraft and the U.S. Navy finally presented realistic development cost estimates to the DOD, the Secretary of Defense himself terminated the A-12 program. Even the DOD's equipment priorities were resource-limited.[6]

For a new airliner such as the MD-12, Douglas would have to seek external investors, either those who saw an opportunity to make a profit where McDonnell did not, or those who were prepared to pay a premium to get into the airliner business with an experienced partner. Such investors might be satisfied with a lower rate of return than John McDonnell or a pure financial trader would accept. Mitsui Bank, whose trading company represented Douglas in Japan, had financed the sale of DC-10 and MD-11 airliners to several carriers in the Pacific. Mitsui Bank expressed interest in investing in the launch of the MD-12 in return for equity, and was prepared to order four aircraft to help launch the venture. There were also extensive discussions with a new company, Taiwan Aerospace Corporation (TAC), in the Republic of China (ROC).

The ROC had for years depended on the United States for its supply of combat aircraft; when U.S. President Nixon extended diplomatic recognition to the People's Republic of China in 1972 and signed the so-called Shanghai Communiqué, one proviso was that the United States would not introduce new and more capable weapon systems to Taiwan but remained free to support existing military equipment (U.S. President Bush breached that provision when he approved the sale to Taiwan of F-16 fighters during his reelection campaign of 1992).

The government of the ROC concluded that it needed to develop its own capability to build combat aircraft, independent of U.S. aid, and launched the development of a new light fighter, powered by

---

[6]In 1998, after a long challenge in federal court, DOD was forced to pay McDonnell–Douglas and its A-12 partner General Dynamics for the costs they had incurred up to the date of cancellation.

two engines derived from Garrett's TFE731 business jet engine by adding an afterburner. In 1978, Garrett had developed the TFE1042-5, a derivative of the TFE731, in partnership with Volvo Flygmotor, the Swedish company. The TFE1042-70 was developed as a substantially new design in partnership with the Taiwanese Aero-Industry Development Center beginning in 1982, but the rationale for U.S. participation remained that it was little more than a relative of the civil TFE731.

The TFE731 was a commercial engine whose export from the United States to Taiwan could be rationalized as being within the letter of the Shanghai agreement. General Dynamics assisted in the design of the fighter's airframe. The engine modification work was to be done by a military research institute in Taiwan with help from the Garrett Engine Company. Many of the Taiwanese scientists and engineers at the institute had design and development experience with American airframe, engine, and electronics firms. Developing a jet fighter and its engine is an expensive matter; hoping that commercial sales could defray some of the costs, the Taiwan government looked for commercial programs that could make use of the facilities funded by the government for development of the so-called Indigenous Defense Fighter (IDF). Taiwan Aerospace Corporation had been set up by the government to make this happen and began discussions with American companies about subcontracts for parts to offset their sales to China Airlines and to the ROC Air Force. Douglas sensed an opportunity to find a partner interested in participating in commercial airliner design and production and capable of funding its portion of such a program through the ROC government. In return for a $1 billion investment toward the development of the MD-12, Douglas offered to merge itself with TAC, with 40% of the equity in Taiwanese hands and 40% of the work to be done in Taiwan. Douglas claimed that Taiwanese participation would enlarge the market for the MD-12 in the Pacific, perhaps even on the mainland in the PRC.

Nothing came of this desperate proposal. The Taiwanese government certainly had the money, but Taiwan had neither the capability to design nor the capacity to build 40% of the MD-12. And the Taiwanese quickly understood that their participation would not suddenly make the MD-12 the hot seller that it had not been before. Taiwan Aerospace Corporation announced that its objective was to be a profitable manufacturer of airframe parts: TAC would be happy to become a subcontractor to Douglas but not a part owner. The conservative Taiwanese gave the same response to British Aerospace when it went looking for investors to keep its BAe 146 airliner in production, and to Dassault when it looked for a partner for a small airliner.

Lacking the possibility to fund development of a new airliner, Douglas became very active in Washington, D.C., claiming that unfair competition from Airbus Industrie jeopardized its continued existence. According to Douglas, the fault lay in European government funding for Airbus development that would never have to be repaid, allowing Airbus to charge lower prices than a private American company could offer, and in Airbus Industrie sales tied to other government agreements such as landing rights. Airbus Industrie reacted with innocent surprise, pointing to the large defense contracts of Douglas and Boeing as evidence of hidden U.S. government subsidies to its own industry (as if the companies participating in Airbus Industrie did not also receive military contracts). Boeing also felt that Airbus Industrie had been competing unfairly but, like the American engine companies whose engines were installed on Airbus Industrie aircraft, was reluctant to enter into a dispute with governments that were also some of its big customers. Boeing was content to have Douglas lead in the trade dispute.

Some ideologues in the U.S. government administration were ready to challenge the European community. The U.S. Congress asked whether it was good for the United States that an American company such as Douglas be driven out of the market by a subsidized foreign competitor, a question with only one politically correct answer. But trade disputes are hardly ever simple: for every industry threatened by foreign competition, another finds a new market. Many American companies were happy to be suppliers to Airbus Industrie, which claimed that about one-third of the value of its aircraft came from the United States. One-third of Airbus Industrie's 35% market share amounted to a greater absolute value of U.S. content than all of Douglas' 6% market share, while Boeing's market share was not noticeably diminished by competition from Airbus Industrie, whether fair or subsidized.

There are several ways of calculating market share: by the number ordered of aircraft of a particular size; the number of seats in the aircraft ordered; or the total price of the aircraft ordered, all in proportion to the total market measured the same way. Effective competition from Airbus Industrie certainly put pressure on price and, therefore, on profit margins but desperate competition from Douglas, with little other than price as a competitive tool, probably exerted equal pressure. After much debate and bluster, the United States and the European community signed the Civil Aircraft Agreement, part of the General Agreement on Tariffs and Trade, both sides binding themselves to buy airliners strictly on the basis of performance, delivery, and price. Both sides agreed not to ask for local production in return for an order, although the seller remains free to offer such "offset" production. Finally, both sides agreed not to tie airliner purchases to other agreements such as those for landing or transit rights. The governments also agreed not to offer sales financing for airliners in the territory of another signatory government and, in general, to limit financing interest rates to some reasonable commercial standard. There was no agreement on subsidies but there was a consensus that government financial assistance for product development ought to be repaid on some reasonable basis out of possible future sales.

The engine manufacturers sat silent during these negotiations, although they no doubt advised their national government delegations of the potential impact of any agreement on their business. Businessmen are notoriously in favor of competition for their suppliers more than for themselves and of subsidies for themselves more than for their competitors.[7] GE and Pratt & Whitney in the United States, Rolls–Royce in England, SNECMA in France, and MTU in Germany all benefited from military contracts that helped generate technology applicable to commercial engines, and some commercial engine programs received national government funding. The engine

---

[7] I am indebted to Murray Weidenbaum, formerly of the President's Council of Economic Advisers, for this insight.

industry saw no advantage in taking sides in an ideological argument over economic principles.

The Douglas product line remained barely alive, helped along by a U.S. Air Force order for 40 KC-10 tanker versions of the DC-10 and a contract to develop and build 80 C-17 transport aircraft for the U.S. Air Force, later increased to 120. The latest Douglas commercial aircraft, the MD-90, another derivative of the venerable DC-9, was to be followed by the MD-95, a shortened version for 100 passengers with BMWRR 715 engines. One could conclude that the "patient" capital invested by the Airbus Industrie governments in the entire product line had taken market share away from Douglas until "milking" the MD-80/MD-90 product line was the only strategy left; at least that was the case Douglas made later to the U.S. trade representative.

Boeing, on the other hand, could afford to react to the competition from Airbus Industrie with caution. It made a number of studies of medium-sized airliners, with two engines, three engines, two aisles or one, some even with double decks, and reviewed them with all its customer airlines. Boeing asked the airlines to calculate the operating costs of such aircraft and suggest what features they would like to see in a new airliner. Boeing stressed the economic advantage of a common airliner design suitable for operation by the largest number of airlines, with maximum flexibility for passengers and cargo in any combination. Until such optimized airliners came along, Boeing told the airlines, continuing to order the 727 was an inexpensive and proven way to add immediate airline capacity. These studies served to slow down any quick commitment to the new Airbus Industrie twin. It was only when airlines stopped ordering the 727 that Boeing seriously offered two completely new twin-engine airliners, the 757 and the 767. The 757 was a narrow body successor to the 727, its fuselage lengthened to carry up to 185 seats six-abreast and requiring two engines of approximately 35,000 lb takeoff thrust each. Rolls–Royce and GE found that they could make such an engine using the cores from their engines for the L-1011 and DC-10 trijet, with smaller fans to fit the 757, whereas Pratt & Whitney designed a completely new engine, the PW2037. This had a new advanced high-pressure compressor and offered lower fuel consumption than either the GE CF6-32[8] or the Rolls–Royce RB.211-535. These, however, had the advantage of proven core engines in service with the airlines and presumably lower development and production costs. And they could be certified and in service much sooner than the new PW2037.

The early 1970s, when Airbus Industrie brought the A300 into the market, was a time of worldwide economic growth. Air traffic was growing and the airlines, fearing that they would lose market share if they did not put capacity in place quickly, were frantically ordering new airliners with lower operating costs, competing for delivery positions with Boeing three years into the future. American carriers were still profitable enough to be able to secure financing for new orders from banks and insurance companies, while European and Asian airlines, most of them government owned, were able to get bank financing with government guarantees as long as they filled their aircraft with passengers, whether profitably or not.

In 1977, however, the equation changed. Fuel prices began to rise dramatically, increasing from about $0.35/gal in January to more than $1.00/gal at the end of the year. A worldwide economic downturn reduced demand for airline travel; airlines began to defer delivery of aircraft they had ordered several years before. Suddenly, lower specific fuel consumption became much more important to the direct operating cost of the 757, outweighing earlier delivery, lower engine price, or the advantage of common logistics from engines with higher fuel consumption. The engine companies took stock. Pratt & Whitney, offering the most modern engine with the lowest specific fuel consumption, announced that it would guarantee that the PW2037 would have a fuel consumption 8% lower than that of any other 757 engine in service. Should that turn out not to be the case, Pratt & Whitney would pay penalties to make good the difference. Delta Airlines recognized a good proposition when it heard one and selected the PW2037 engine for its 757s.

---

[8]Around 1974, GE's Advanced Product Planning Operation proposed that Boeing certify a version of the 727 with two CF6-32 engines. Boeing ran some wind-tunnel tests but saw no incentive to going further as long as airlines continued to order the successful base-line design with three JT8D engines.

Rolls–Royce, then still owned by the British government, decided to continue to fight for a share of the 757 market, confident that it would not be allowed to go out of business, no matter how unprofitable its transactions. Rolls–Royce also speedily improved the RB.211-535 so that Pratt & Whitney almost certainly had to pay a fuel premium to Delta Airlines. With Pratt & Whitney having the better engine and Rolls–Royce determined to compete at any cost, GE concluded that this was a poker game it could not win and stopped the CF6-32 program. Rolls–Royce stayed in and has taken a share of about half of the 757 market with its RB.211-535, which has turned out to be a very efficient and reliable product.

The PW2037 has also been successful in the 757 and has been ordered by the U.S. Air Force for the C-17 transport aircraft and by the Russian airframe firm Ilyushin for the Il 96. So far, the Il 96 program has received 10 flight test engines from Pratt & Whitney, as Pratt & Whitney's investment in developing the Russian market. In December 1996, Aeroflot announced that it would order 20 Il 96-300 airliners with Pratt & Whitney engines and Honeywell avionics, 7 with their price guaranteed by the Russian Government. The other 13 have a much more modest guarantee and there is some question whether they will be built at all.

Boeing was aiming at competition posed by the Airbus Industrie A300 and its even more advanced A310 derivative. The 757, the direct successor to the 727, was only one part of the Boeing strategy. Its direct response to the A300 was another, larger aircraft, the 200 to 250 passenger 767 with twin aisles and two engines. For this, Pratt & Whitney offered a larger version of the new PW2037 design, the PW4000. GE also had an engine of the right size and performance, the CF6-50, but to compete with the new PW4000, the CF6-50 design needed to be modernized, made lighter and simpler. For the new CF6-80A, GE retained the compressor and turbine aerodynamics but designed a completely new support structure for the compressor and turbine rotors, eliminating an entire structural frame behind the high-pressure turbine. The GE1/6 technology demonstrator that GE ran in 1964 to validate the concept of a bypass ratio much higher than seen before had no interturbine frame. But GE's engineering conservatism led it to incorporate such an interturbine frame in the TF39 and CF6-6 and -50, only to get rid of it again in the CF6-80A.

The first major engine competition for the 767 came at United Airlines (UAL), a critical launch customer for both engine companies. After a vigorous campaign, United picked the PW4000, in part because it believed that the CF6-80A had limited potential for growth. A competition at American Airlines followed immediately and was won by the CF6-80A at no doubt considerable cost to GE. GE did whatever it took to make the CF6-80 attractive to American Airlines and won the order.

To win a competition at a reasonable price, the product must be better technically, and that better quality must have a measurable impact on the airline's operating costs. For long-term success, GE recognized that it needed a better technical solution. GE designed a new and larger fan and certified the engine as the CF6-80C for the 767 and for use in the new Airbus Industrie twin, the A310. The new engine had a rated thrust up to 60,000 lb—large enough to serve the 747 and the MD-11. Later GE fitted it with an even larger fan for the even larger Airbus Industrie A330. For the 767 and the Airbus Industrie A310 the new engine was a perfect match and solved the problem of limited growth capability raised by United. It had more thrust and better fuel efficiency than its competitors. As the results came in from airlines using the engine, it became a brilliant commercial success for GE.

Chapter 8

The Military Market Share

GE's early dominance of the U.S. jet engine industry had been lost to Pratt & Whitney, whose dual-rotor military engines and their commercial derivatives found growing market acceptance. Income from these products gave Pratt & Whitney the resources to launch several completely new series of commercial engines without military precursors—first, the JT9D for the 747 and, later, the PW2037 and PW4000. By the mid-1970s, GE had less than a quarter of the world commercial engine market and about a quarter of the U.S. government's military engine business. It had successfully developed and qualified the F101 engine, a very advanced afterburning dual-rotor turbofan for the supersonic B-1 bomber and delivered 44 of them for aircraft flight testing when U.S. President Carter cancelled the B-1 bomber program (the program would be reinstated some years later by Reagan's administration). That left GE with the dual-rotor TF39 engine in production for the Lockheed C-5A, the J85 engine for the Northrop T-38 trainer and F-5 fighter, and the J79 engine—which would stay in production for a few more years, with a worldwide demand for spare parts—as well as the small helicopter turboshaft engines produced in Lynn.

Northrop had converted the T-38, its successful supersonic trainer for the U.S. Air Force, into a lightweight fighter, the F-5, using GE's J85 engines in both. The F-5 had sprightly performance but was limited in range and payload. It was inexpensive as supersonic fighters go because aircraft cost is roughly proportional to airframe weight, and high production of the T-38 had brought costs down. Northrop advocated the F-5 as the perfect machine for the air forces of a number of countries allied to the United States.

The United States was not alone in developing military aircraft during the Cold War. The former Soviet Union designed and built combat aircraft for its own air force and those of its satellites, unveiling new aircraft types of greater performance much more frequently than the NATO alliance; Britain and France had a long history of design and production of military aircraft, which both countries protected jealously; Germany and Italy had a lesser independent capability after the end of World War II but continued to support a degree of design capacity and manufacturing resources commensurate with their need for aircraft. To defend NATO effectively against a potential attack by the Soviet Union, the United States encouraged standardization of aircraft and other weapons to the extent that each member country's sponsorship of its national industry would allow, and encouraged the major NATO countries to use their own funds to equip themselves. That was why the United States liberally made technical data available to its NATO allies for their licensed production of major weapon systems. (Standardization and rationalization in the Warsaw Pact were simpler: all member countries used Soviet equipment; major weapon systems were built in the USSR, and others were assigned one to a satellite country for production.) But there were many countries allied with the United States that, by necessity, depended on U.S. government financing for their fighter aircraft procurement, some of them through outright grants, others through the use of the U.S. DOD as their purchasing agent, more expert and less corruptible than any of their own. For example, the Federal Republic of Germany received Canadair Sabre fighters (a version of the F-86) and later F-84 fighters as grant aid from the United States, before Germany began to invest its own funds in the licensed production of the F-104 Starfighter. Grant aid also paid for fighters for Pakistan, Turkey, Greece, Morocco, Taiwan, and Spain and many other countries strategically important during the Cold War but without adequate resources to pay for their own aircraft. This had been U.S. policy since the beginning of the Cold War and had led to large exports of U.S. military aircraft. In principle, President Carter and a sizeable political constituency in the United States opposed such exports of U.S. fighter aircraft, arguing that they led to arms races among poor countries, diverting money better used for social purposes; indeed, the United States itself had better uses for funds spent on grant aid and foreign military sales programs. Many of the recipient countries were not reliable politically. They might use U.S. arms aid against each other and, in the worst case, even against the United States itself. But others in the U.S. government believed that there was political value to equipping allies to defend themselves and participate in mutual defense. Control over the supply of spare parts, ammunition, missiles, and bombs gave the United States some control over how such aircraft could be used. Each export sale received careful political scrutiny in the U.S. Departments of Defense and State and the U.S. National Security Council, and the risk could be managed. (Thus, Greece and Turkey never used their F-104s and F-5s against each other during their periodic face-offs in the Aegean.)

The aircraft firms liked military export sales, whose profit margins were often higher than in sales to the U.S. government for its own use. They argued that increased production reduced the unit cost to the U.S. government and kept American production lines busy at times of reduced U.S. demand. Foreign governments would find ways of satisfying their desire for fighter aircraft from other suppliers if the United States held back. The Soviet Union, Great Britain, and France were usually ready to supply military equipment to such markets. France in particular was an aggressive seller of fighters. The French government nurtured an indigenous capability for the design and production of combat aircraft, in part because General de Gaulle was unwilling to depend on "the Anglo-Saxons" as suppliers and, in part because of industrial policy. The French government saw its aeronautical industry as an instrument of national economic policy, a source of exports, a motor of industrial technology for other French industry, and as a tool in the competition for leadership in Europe with the Federal Republic of Germany.

Typically, the French air force ordered 100 of a new fighter design from Dassault, the sole French firm involved in fighter production. A French government agency responsible for exports of military equipment marketed the new design around the world. When an export order was received, the government export agency was able to promise quick delivery out of the initial batch ordered for the French air force, which then ordered replacement aircraft for itself. Export orders increased production quantities and reduced the costs of France's own military procurement. Israel, Switzerland, Belgium, and a number of South American countries bought French fighters. Through this ratcheting process, France was able to keep up the development of its own military aircraft longer than Britain, for example.

Affordability is always a major issue in the development of new fighter aircraft. In 1957, Duncan Sandys, the British Minister of Defence, concluded that Britain could not afford to continue its independent combat aircraft development, and effectively changed British policy to the procurement of some aircraft and weapons from the United States and to participation in European cooperative programs for aircraft development and production, such as the subsequent Jaguar and Tornado. But it was not until some 40 years later, in 1996, the French government was finally forced to admit that it might not be able to maintain its strategy of complete independence for new equipment.

NATO accepted West Germany as a member in 1957 and agreed to the equipment of the German air force with modern jet strike fighters. As an interim step, the United States furnished F-84 and F-86 fighters under grant aid. The political plan was that, in the long run, Germany would use German money and German production to equip the German air force with modern aircraft for the common defense against the former Soviet threat—but always as part of the NATO command, not as an independent military force. The Germans considered several fighter aircraft: the British Lightning, the French Mirage III, the American Grumman F11F-1F, and the Lockheed F-104. Another American aircraft, the Republic F-105, was withheld from the competition by the U.S. government, which deemed its range and payload capability too great. The F-105 could reach Warsaw and even further east with a significant bomb load. That was fine for the U.S. Air Force in Europe but politically unacceptable for the German air force: German operation of the F-105 would be perceived as a provocative threat to the USSR.

In 1958, Germany selected the Lockheed F-104G Starfighter and launched a program for the production of more than 800 aircraft. Other NATO countries saw the military and financial advantages of standardization, and the opportunity for participating in a giant production program. Soon Canada, Belgium, the Netherlands, and Italy also adopted the F-104 as their next fighter. Japan followed suit. In subsequent years, the U.S. government would sell or give F-104 fighters to Norway, Denmark, Greece, Turkey, Jordan, Pakistan, and Taiwan. The F-104 was put into production in Germany, Holland, Belgium, Italy, Canada, and Japan, and its J79 engine was produced in all of these countries except Holland. Assembly lines were set up in several countries but production of each part was typically centered in one, which supplied all of the assembly lines. The existing production sources in the United States, at Lockheed and GE and their American suppliers, backed up the international program. It was a gigantic industrial enterprise, the first Mach 2 fighter and its engine produced in most of these countries.[9] With technical assistance from Lockheed, GE, North American–Rockwell, and Litton, more than 2000 Starfighters and their equipment were built, all successfully meeting the same standards of performance and quality.

In every country, the selection of fighter aircraft is a complicated mixture of tactical requirements, affordability, and industrial policy. Thus, when production of the F-104G ended in Italy, the Italian government wanted to find a way to maintain employment in its aviation industry, some of it newly set up in the impoverished south. Any new aircraft had to be an improvement over the F-104G, of course, to justify putting it into production at all, but too big an improvement would not be affordable. The Italian government de-

---

[9]Japan had built pulse-jet powered missiles at the end of World War II. Germany had the Me262 fighter in large-scale production at the end of the war and two axial flow turbojet engines, the Junkers 004 and the BMW 003. After the war, Belgium built British and French jet engines under license. For a while Canada continued with aircraft and engine development on its own. When Canada adopted the F-104, it canceled development of the Avro Arrow fighter and the Orenda Iroquois engine. To the dismay of Orenda's engineers, they were informed a week later that they would begin to build the GE J79 engine under license.

cided to fund the development of the F-104S with Lockheed, and its engine with GE. The F-104S had an advanced radar, a more powerful engine, a bigger wing, and could carry two Sparrow missiles, giving it better intercept capability.

In Washington, meanwhile, Northrop was advocating the F-5 as the perfect replacement for the F-104. It was less expensive and had enough supersonic performance to satisfy any head of state. Besides, its limited payload and range gave its operators less opportunity for mischief against the interests of the United States. These were persuasive arguments, and F-5s soon appeared in the air forces of Canada, Brazil, Chile, Venezuela, Thailand, Taiwan, South Korea, the Philippines, Indonesia, Saudi Arabia, Turkey, Greece, Morocco, Norway, the Netherlands, and Switzerland. The airframe and engines were also put into production in many of these countries. The F-5 program became another giant international production program, remarkable for the fact that almost all of the aircraft were built for use outside the United States. While the U.S. Air Force used F-104 fighters for a number of years in Air Defense Command and Tactical Air Command, the only F-5 fighters in the U.S. Air Force were used to simulate former Soviet fighters for air-to-air tactical training.

Of the European F-104 users, Norway, Spain, Greece, Turkey, and the Netherlands adopted the F-5. Britain, Italy, and Germany, reasoning that their combined requirement for high performance strike fighters was big enough to justify the cost of a European design, developed the Tornado, a twin-engine swing-wing Mach 2 strike aircraft, and its Rolls–Royce RB.199 engine. In addition, Britain developed a fighter version of the Tornado. They had visions of exporting Tornado aircraft to other countries. In the case of the F-104, the United States paid for the design and the Europeans needed to pay only for its cost of production in Europe. Now the Tornado syndicate faced the need to pay for the entire development as well as the production program. The Tornado is a complicated and capable combat aircraft and, as is the norm for all such devices, its development took far more time and cost far more money than participating governments had expected when they gave their approval. The Tornado was not going to be the world best seller that the F-104 and the F-5 had been: it was capable but too expensive even for its original developers.

That was not a problem limited to Europe. The U.S. Air Force ordered the McDonnell F-15 fighter and its Pratt & Whitney F100 engines as the replacement for its F-4 Phantoms but, like the Tornado, the F-15 was much more expensive than had been forecast when the U.S. Congress approved the launch of the program. Clearly the U.S. Air Force would not be able to buy all of the F-15s it wanted for its tactical fighter force. Given the strict budgetary constraints felt by the wealthy United States, one might wonder how the USSR could afford to develop a steady stream of new advanced fighters and put them into mass production. In the end, of course, such military procurement was one of the causes of the Soviet Union's financial collapse.

The U.S. Air Force's answer was a competition for a "lightweight fighter," under a new set of rules: instead of its normal massively detailed specifications, the U.S. Air Force set simple requirements for aircraft price, speed, payload, range, and maneuverability. How these were met was up to the airframe designers. The U.S. Air Force expected to buy at least 1000 of the light-weight fighters to complement its fleet of F-15s.

The U.S. Navy paid for the development of a different aircraft, the Grumman F-14, to replace its F-4 Phantoms, with the Pratt & Whitney TF30 engine first developed for the F-111, but the TF30 was not powerful enough to give the U.S. Navy the performance it wanted from the F-14. The U.S. Navy then paid Pratt & Whitney to develop the F401, a more powerful version of the U.S. Air Force's F100. Development was terminated after several major failures. Years later, the F-14 received more powerful GE F110 engines and finally achieved the performance specified by the Navy from the beginning.

Of the European F-104 users, Germany and Italy had committed themselves to the Tornado. The others, concluding that this U.S. Air Force lightweight fighter made sense for them too, joined the competition with the commitment from the United States that they would make 40% of any lightweight fighter they bought for themselves and 30% of any lightweight fighters sold to other countries. The balance of the hardware would come from the United States.

Two airframe companies entered the U.S. Air Force competition, with radically different designs. General Dynamics designed a single-engine fighter using the same Pratt & Whitney F100 engine already installed in the F-15, a concept favored by the U.S. Air Force. A standard engine for the F-15 and the lightweight fighter offered obvious logistics advantages. Increased production of the F100 would also bring down its costs and indirectly make the F-15 more affordable.

Northrop designed a twin engine fighter with a completely new engine from GE, the YJ101. Tom Jones, the chairman of Northrop, had always believed that its T-38 two-seat supersonic trainer powered by two GE J85 engines could be turned into a relatively inexpensive supersonic fighter that could then be marketed to all of the air forces equipped with F-84, F-86, and even F-104 aircraft under U.S. grant aid programs. Northrop modified a T-38 into a single-seat prototype lightweight fighter. The U.S. Air Force gave its blessing and the type designation F-5A. Northrop built several thousand and sold them to foreign air forces, which unlike the U.S Air Force could not afford to buy and operate larger fighters such as the F-4. Some F-5s were sold or given by the U.S. Air Force to air forces that were, on occasion, tempted to use them against other allies of the United States, for example, Greece and Turkey, then and now embroiled in a dispute over Cyprus and jurisdiction over mineral rights in the Aegean, or Jordan and Saudi Arabia, both at the time technically at war with Israel. The limited range and payload of the F-5A was a political advantage in such cases.

Edward Woll, who had led GE's J85 program for the T-38 and F-5A, conceived a growth version of the J85 whose higher airflow would improve the performance of the F-5. GE's flight test center at Edwards Air Force Base in California modified an F-5A to accept the more powerful engine, and a GE test pilot demonstrated the improved aircraft performance (in its classically self-confident style, the Royal Canadian Air Force developed another model of the J85 for its own F-5s, with performance between that of the J85 in the F-5A and the more powerful J85 in the F-5E. GE assisted the Canadian engine development program.) Northrop at once took up the idea and successfully campaigned for U.S. Air Force support of the new F-5E as the fighter for U.S. international allies. When the U.S. Senate investigated illegal corporate contributions by Northrop to the Committee to Re-Elect President Nixon, it found evidence of large cash payments by Northrop as commissions to middlemen who facilitated the sale of F-5A and F-5E aircraft to the governments of Iran and Saudi Arabia. Northrop protested that it was merely following the marketing practice of the region, which Lockheed had also used before it in the sale of F-104 fighters and C-130 freighters. Prominent figures in the governments of Germany, the Netherlands, Iran, and Saudi Arabia were said to have received large sums of money. Such practices were not acceptable to the U.S. electorate in the aftermath of the Watergate investigation, and these revelations led to the Foreign Corrupt Practices Act of 1977, which makes it a federal crime for an American company to pay bribes to an official of a foreign government to secure a sale.

Woll had seen the need for a more advanced engine cycle for the lightweight fighter than that of the single-rotor J85. The YJ101 was a radical innovation for GE, a dual-rotor turbofan (like Pratt & Whitney's TF30 and F100), but with a much lower bypass ratio and a much simpler mechanical structure. The YJ101 cycle proved to be brilliantly adapted to fighters. The bypass ratio was so low that the U.S. Air Force Propulsion Laboratory joked about the YJ101 as the "leaky turbojet." The dual rotor compressor had very high performance and efficiency and, at the same time, was extremely stable during power changes at altitude and during violent aircraft maneuvers. The simple mechanical structure gave the engine low weight and promised low cost in production.

General Dynamics and Northrop engaged in an aggressive marketing campaign, both to the U.S. Air Force and to the European air forces. Both lightweight fighter aircraft demonstrated very high performance and also appeared to satisfy the cost objectives. In the end, the U.S. Air Force selected the General Dynamics design, the F-16, in part at least because it used the same engine as the F-15. Holland, Belgium, Norway, and Denmark followed the lead of the U.S. Air Force, and the F-16 and its engine were put into production in those countries, in a program analogous to the F-104 production program of 20 years before.

Northrop did not concede defeat. The U.S. Navy had not adopted the F-16 and was known to prefer twin-engine fighters. Northrop allied itself with McDonnell–Douglas, a traditional supplier of fighters to the U.S. Navy, and transformed its lightweight fighter into the F-18, using an evolution of the YJ101 engine developed by GE. The F404 engine had the same compressors and simple mechanical design, but a slightly higher bypass ratio and more thrust. The F-18 was adopted by the U.S. Navy and put into large-scale production, with McDonnell as the lead contractor.

Other air forces liked the presumed safety factor of two engines. The F-18 was adopted by the air forces of Switzerland, Canada, Australia, and Finland, with partial production in all those countries. With its excellent performance and stability, the F404 engine was adopted by Sweden for its Gripen fighter, by Singapore for its modified A-4 strike aircraft, and by India for its light combat aircraft. Israel had been designing a light strike fighter called the Lavi and had considered the GE F404 as well as a Pratt & Whitney engine. But like most other countries, Israel found itself running out of money to meet all the claims on its military budget. When the U.S. government made F-15 and F-16 fighters and military aid funding available, Israel dropped the development of the Lavi and adopted the American fighters. Turkey adopted the F-16, as later did Egypt, Greece, the Republic of Korea, Singapore, and Pakistan. But there was a new wrinkle: the F-16 had become available with another engine, the GE F110.

Pratt & Whitney's F100 engine had exhibited a tendency to compressor stall when pilots called for more power at altitude and during extreme maneuvers, a distressing repetition of the problems of the Pratt & Whitney TF30 in the F-111. F100 compressor stalls occurred so frequently that the U.S. Air Force was forced to adopt temporary restrictions on the speed and altitude of its F-15 fighters while Pratt & Whitney worked to correct the trouble. In a single-engine fighter such as the F-16, compressor stalls were simply unacceptable. Pratt & Whitney complained that its engine was being blamed for not meeting new requirements that were stiffer than those specified when the F100 was being developed. No matter where the merits of that argument lay, the U.S. Air Force was dissatisfied with its F100 engines and Pratt & Whitney's attitude to solving them. As a way to put pressure on Pratt & Whitney, and perhaps as an insurance policy to back up its corrective action program, the U.S. Air Force sponsored GE's development of the F101 derivative fighter engine.

The F101 had earlier been qualified by GE for the B-1 program and was again put into production when President Reagan reinstated the bomber program. The 44 F101 engines GE had delivered for the original B-1 flight test program performed flawlessly. Woll saw that by changing the fan to one of a lower bypass ratio, and by replacing the F101 afterburner with another scaled up from that of the F404, simpler than the afterburner required for the B-1 bomber, GE could turn the F101 into an engine suitable for high-performance fighters. Woll committed GE to building a prototype of such an F101 derivative fighter engine, which looked much like a scaled-up F404. After successful ground and flight tests, GE received support from the U.S. Air Force for continued development. The F101 derivative fighter engine flew successfully in several versions of the F-16 and exhibited the same outstandingly stable compressor performance as the F404. It also had more thrust than the F100. If the U.S. Air Force chose to use it, it would solve its operating problems with the F100.

The U.S. Air Force decided to hold a competition for production of F-16 engines. Thrust was not an issue because the F-16 was already deemed to have acceptable performance with the Pratt & Whitney F100 engine. But operating reliability and stability were the key criteria, and engine price was important, as well as life-cycle costs over years of operation. Pratt & Whitney set to work on the problems of the F100, with support from the U.S. Air Force, which after all depended on the engine for its frontline F-15 and F-16 fighters. Pratt & Whitney also tried to persuade the DOD and U.S. Congress that it was a mistake for U.S. Air Force to consider an alternate engine for the F-16. Robert Carlson, the Pratt & Whitney president, told a Congressional committee that "it would be stupid to change the engine of the F-16." No doubt the U.S. Air Force did not relish being called stupid. After it evaluated both designs, it concluded that both the GE engine, now designated the F110, and the improved

Pratt & Whitney F100, were acceptable technically. The higher thrust of the F110 was never officially a factor, although U.S. Air Force pilots would, during the Gulf War, be able to take advantage of it by flying above Iraqi antiaircraft fire with a full bomb load, something they could not have done with the F100 engine.

To Pratt & Whitney's immense surprise and chagrin, and even though it had by then delivered more than 2000 F100 engines to the F-15 program, the U.S. Air Force selected GE based on a better price. GE's price was lower because the F110's mechanical design was simpler than that of the F100 and because of the beneficial learning-curve effects of the production of F101 engines for the B-1 and CFM56 engines for other aircraft (see Ch. 12 for more detail). All three engines shared the same core. Perhaps also Pratt & Whitney had become a little complacent after many years of dominating the U.S. Air Force fighter engine market without effective competition. The U.S. Air Force thereafter split its F-16 engine purchases almost evenly between Pratt & Whitney and GE, retaining the improved Pratt & Whitney F100 for the F-15.

Higher thrust clearly was an important factor when the U.S. Navy adopted the F110 to replace the much less powerful Pratt & Whitney TF30 in its F-14s; it had already selected the GE F404 for its F-18s. At this point, the contractual process took a surprising turn, one that would have far-reaching effects. John Lehmann, Secretary of the Navy, concerned at GE's sudden dominance as the U.S. Navy's engine supplier and angry at what he saw as GE's intransigence during contract negotiations, awarded a contract to Pratt & Whitney for 100 F404 engines and directed GE to send drawings and two complete engines to its fiercest competitor. GE and many others in the industry were just as upset at what they perceived as the U.S. Navy overreaching its legitimate contract rights. Feelings ran high in both factories. In earlier competitions, Pratt & Whitney's assembly line in Hartford, Connecticut, had been plastered with posters with pictures of Gerhard Neumann with the caption "This man wants your job!" GE's workers in Lynn were so incensed at being asked to help their competitor that they walked out on a two-day wildcat strike against the U.S. Navy.

Pratt & Whitney's cost for the engines was understandably high, reflecting the start-up of production. It was a perverse situation. The

U.S. Navy agreed to reimburse the cost to Pratt & Whitney, with a very tight allowance for overhead and profit, using GE's lower price as a negotiating tool. It was a perverse situation. The U.S. Navy agreed to pay a higher price to Pratt & Whitney to build an engine designed by GE. In reaction, GE took a brave step into the future: it proposed an attractive low fixed price for the F404, lower than any the U.S. Navy had paid so far—provided the U.S. Navy agreed to "buy out" the program from GE, in other words, to give GE an order for all of the F404 engines it would buy. The price was to be a fixed one, subject only to economic escalation for factors outside GE's and the U.S. Navy's immediate control.

Secretary Lehmann agreed. Pratt & Whitney was now a second source for the F404, able to step in if GE were unable to deliver engines, and from GE the U.S. Navy had a fixed price lower than it had been ready to pay before the competition. As GE's Krebs said at the time, "There's nothing like production volume and multi-year fixed price contracts to drive costs down." Competition also clears the mind powerfully. The military business was beginning to take on more and more of the trappings of the commercial engine business.

As soon as the GE F110 became available, the Israeli air force selected it for its F-16s.[10] Unlike the U.S. Air Force, the Israelis put great stock on improved performance because their enemies were only a few minutes flying time away, and the F110 had more thrust than the Pratt & Whitney F100. Perhaps there was also some dissatisfaction with Pratt & Whitney's responsiveness to F100 problems. Like the U.S. Air Force, the Turkish air force originally picked the F-16 with its Pratt & Whitney engine for its next fighter but, after the U.S. Air Force and the air force of Israel made their choice of the GE engine, Turkey reconsidered and also changed to the F110.

Greece chose the F-18 but, later amid cries of bribery, switched to the F-16 and, after an engine competition, picked the F110 as Turkey had done. One potential antagonist affects the other. Israel se-

---

[10]The sale had a tragic personal consequence. A GE employee reported to the U.S. government that a GE manager "skimmed" funds from the contract in a conspiracy with the chief of procurement of the Israel air force. The American and the Israeli, both highly regarded until then, were sent to jail.

lected the GE engine in the F-16. Egypt did the same. On the other hand, Pakistan, Singapore, and Korea picked the Pratt & Whitney engine for their F-16s. Like Greece, South Korea had first chosen the F-18 and then changed the selection to the F-16 amid allegations that it had been bribed by General Dynamics. No evidence was ever found to substantiate the allegations, neither in Korea nor in Greece. Korea emphasized that the Pratt & Whitney F100 engine it selected was installed in half of the U.S. Air Force F-16s and was common to the engines of U.S. Air Force F-15s stationed in Korea and Japan. That was good enough for Korea.

As always, Japan went its own way. The Japanese had built the F-104 under license, followed by the F-4 and the F-15. Along the way, Mitsubishi Heavy Industries led a consortium of airframe companies to design and build the T-2 trainer and its F-1 attack version, Japanese re-creations of the Anglo–French Jaguar, using the Jaguar's Adour engines designed by Turbomeca in France. As the Japanese industry gained experience in the manufacture of modern combat aircraft, it began to lobby for the design of an indigenous fighter to succeed the F-1 and to complement the F-15. The technocrats in the Technical Research & Development Institute (TRDI) of the Defense Agency and the officials of the Ministry of International Trade & Industry (MITI) supported the plan. Flight safety was of paramount importance in Japan, and the technocrats decided that only a twin engine fighter design would be acceptable. Not surprisingly, Mitsubishi's design studies looked a lot like the McDonnell/Northrop F-18. The Technical Research & Development Institute even decided on the GE F404 engine and set in train the long process of approval, the *nemawashi* or "binding of roots" that would result in a consensus to be ratified by a formal cabinet decision. But politics intervened in a military expansion of the trade disputes between Japan and the United States. In Japan, the airframe industry and the Defense Agency wanted to design and build a Japanese fighter to demonstrate Japan's maturity and emergence from the shadow of building American fighters under license. Opposing this, the Ministries of Finance and Foreign Affairs wanted to maintain the U.S. defense umbrella of the Mutual Security Treaty and to avoid inflaming more U.S. complaints about the trade imbalance between the two countries. If

Japan insisted on designing its own fighter and refused to import an American product in the one industrial sector where the Americans had an undoubted superiority in product design and cost, that would play into the hands of those in the United States who threatened economic retaliation against Japan's closed markets.

There were several factions in the United States also. The Defense Department wanted Japan to take a more assertive role in support of the Mutual Security Treaty and pushed the Japan Defense Agency toward extending the Japanese defense perimeter to 1000 miles from the home islands. Such a strategy would require the purchase of longer-range airborne early warning (AEW) aircraft to supplement the shorter-range Grumman E-2C aircraft forced on Japan by President Nixon in 1972 (there were allegations at the time of a "black rain" of bribes flowing from Grumman through its trading company Nissho-Iwai to the Liberal Democratic party of Prime Minister Tanaka). The Japanese purchase of an expensive AEW system, obtainable only in the United States, would also have a huge and beneficial effect on the U.S.–Japan balance of payments.[11] If Japan were to design its own fighter, it would need an American engine and other technical assistance. In the DOD's view, that was an acceptable concession in support of the overall joint defense strategy. The State Department, however, had reservations about an extension of the Japanese defense perimeter. Most of the other countries in East Asia viewed the concept with some alarm, remembering Japan's military hegemony before and during World War II. Korea and China were particularly hostile to the notion.

The U.S. president's National Security Council understood these reservations but did not object to Japan's desire to go ahead with its own fighter. Engine development takes longer than airframe development and costs as much. Lacking their own program to develop an indigenous fighter engine, the Japanese would have to use an American engine and probably would also need assistance in avionics.

---

[11]In the end, Japan ordered an airborne early warning and control system based on the Boeing 767 (used by JAL and ANA) and the Westinghouse AP-142 radar. The American AWACS based on the Boeing 707 had gone out of production by the time the Japan Defense Agency made up its mind.

Cooperation would give the United States insight to what the Japanese were doing and access to any unique Japanese technology. It would also enmesh the United States so closely in the Japanese program that it would restrict Japan's freedom for independent action: divergence in foreign policy, for example, or disputes in trade policy. Balancing all of the advantages and disadvantages, the U.S. government under Presidents Reagan and Bush was supportive of the Japanese desire to design the Fighter-Support, Experimental (FSX).

Many in the U.S. Congress did not agree. The trade imbalance between the two countries had become a matter of hot debate, fanned by a U.S. automotive industry hard pressed by competition from Japanese imports. Chrysler chairman Lee Iaccoca became a prominent and effective spokesman against Japan and its restrictive practices against imports. That Chrysler had no small cars suited to Japanese roads, none with a fit and finish competitive with that of Japanese vehicles, and none with the steering wheel on the right, did not deter Iaccoca for a minute. The real issue for him, of course, was not that Japan would not import American cars, but that it insisted on exporting Japanese cars to the United States, where perverse consumers insisted on buying them.

The U.S. Congress made a sensible point: why should the United States concede its dominance in aviation, the one economic sector where it continued to lead the world? Acceding to Japan's desire to design its own fighter would be a mistake. If Japan insisted, we might not be able to stop it but we should then make Japan pay in some other way. To this school of thought, helping a Japanese program with American technology was suicidal, a replication in aviation of the same sad surrender of industrial supremacy that had already taken place in consumer electronics, steel, and automobiles. Many Americans argued that Japan could import a fighter such as the F-18 for a third of what it would cost to build under license in Japan; designing and building a purely Japanese fighter would cost even more than that because production quantities were bound to be small. Nor could there be offsetting profits from sales abroad, because Article IX of the Japanese constitution more or less prohibited the export of military equipment, limiting FSX sales to the Japan Defense Agency.

The Japanese argued that licensed manufacture in Japan was actually cheaper. Under life-time employment, aircraft factory workers would be paid whether they did useful work or not. The only choice to the Japanese was between importing complete aircraft or only the raw material for them. Obviously the latter cost less.

Such debates are seldom models of reason or a calm weighing of arguments. The dialogue became increasingly ill-tempered and racist on both sides of the Pacific Ocean. There was also a sharp contrast between public statement and real meaning, a technique the Japanese have practiced for years as *tatemae* and *honne*. When U.S. Defense Secretary Caspar Weinberger announced that making an American engine available for the FSX had nothing to do with trade policy, only with mutual defense, nobody in Japan believed him for a moment and those Americans who half-believed him were appalled. But had the U.S. National Security Council made public its subtle calculations around the advantages of cooperation between Japan and the United States, they would have been derided at home and would have caused a storm of protest in Japan.

Every angry argument in America produced its angry response in Japan. Perhaps in the end the process was constructive, demonstrating to the political public in both countries that there was strong opposition on both sides to the nationalist position and that it was in the interest of both countries to reach an accommodation balancing the issues of mutual defense and trade. The Japanese cabinet decided to go ahead with the design of the FSX. Unlike the configuration originally chosen by TRDI, which resembled the F-18, it now looked much more like a development of the F-16! Reluctantly, TRDI agreed to American participation in FSX development and the sharing of advanced Japanese technology with the United States. That is not an easy process. Fifty years after Japan had had the reputation of manufacturing shoddy products, TRDI was now convinced that certain Japanese technologies were superior to those in the United States. For example, Japan led in the manufacture of large monolithic structures made of graphite fiber and plastic resin, and in the manufacture and application of gallium–arsenide semiconductors. The Japan Defense Agency maintains that intellectual property rights reside with the Japanese firms and cannot be ceded by the government. In fact, the U.S. military airframe industry was not very open to the notion that Japan might have superior product design or manufacturing technology, although Boeing has subsequently taken advantage of this in the case of 777 production in Japan.

The F100 engine, which Japan was already building under Pratt & Whitney license for the Japanese F-15, would have been an easy choice for the FSX. However, after an intense competition, TRDI selected the GE F110 engine for its new FSX fighter, perhaps to take advantage of the higher thrust and reliability or—just as likely—to gain access to yet another source of engine technology.

On the heels of their Tornado cooperation, Germany and Britain decided to design their own new fighter, the Eurofighter, which was to be very agile and with very high performance, just like every other new fighter. Again Italy joined the program, as did Spain later. Following the pattern established in the Tornado program, decision-making was slow, and there was great skepticism about program cost estimates. On several occasions, the German parliament threatened to deny funds for development, let alone for production. One problem was that the collapse of the Soviet Union reduced the urgency of the need for a new fighter and also put first-class Soviet aircraft on the market, including several squadrons of MiG 29 fighters from the former East German air force. One squadron of MiG 29s was kept in service by the German air force for comparison testing against the F-16, F-18, and Eurofighter. Three more squadrons were bought by the U.S. government from Moldova to keep them out of the hands of Iran.

The program for the Eurofighter and its new engine, the VJ200, was finally launched. An American cynic commented that the Europeans would only succeed in producing an improved F-18, at enormous cost and with a delay of 10 years—without endowing it with any stealth capability. Meanwhile in France, Dassault launched the development of the Rafale, another Eurofighter/F-18 look-alike. Until SNECMA finished developing its own M88 engine for the Rafale and produced engines for flight test, the Rafale had to make do with GE F404 engines, which the M88 greatly resembled. While Japan was developing an improvement of the F-16 and the Europeans what might be called incremental improvements on the concept of the F-18, in the United States the cycle continued of developing quantum advances in performance over aircraft in service. One such advance was the F-22 Advanced Tactical Fighter, the replacement for the 20-year-old design of the F-15. The new Advanced Tactical Fighter would be "stealthy" and capable of supersonic cruise without using its afterburners, to make it more difficult to detect by radar or infrared missiles. The U.S. Air Force selected the new Pratt & Whitney F119 engine design for the F-22. GE's F120 engine was adjudged more advanced but also more risky. The F119 met the U.S. Air Force's performance requirements, and that was good enough.

Just as the F-16 became the inevitable low-cost complement to the F-15, the F-22 would also need a less capable but less expensive complement, the new "Joint Strike Fighter" (JSF). The JSF would succeed the F-16 in the U.S. Air Force; the AV-8B VTOL fighter in the U.S. Marine Corps; and the Harrier in the British Royal Navy. Like the F-16, the JSF is to be single-engined and, like the F-22, "stealthy." The Pratt & Whitney F119 and the GE F120 are both in contention for the JSF. For the V/STOL version for the U.S. Marine Corps, GE has allied itself with Rolls–Royce and Allison, Rolls–Royce's American subsidiary, because Rolls–Royce has the most operating experience with VTOL engines. British Aerospace (BAe) is a partner of McDonnell Aircraft, now merged into Boeing and one of the two airframe competitors for the JSF; BAe is also talking with Lockheed–Martin, the other competitor. No matter who wins the competition, BAe expects to participate in proportion to the British procurement of the JSF and in recognition of British expertise in VTOL aircraft and engines.

In this continuing back-and-forth battle for market share, GE managed to gain more than half of the U.S. military engine market at the expense of Pratt & Whitney between the mid-1970s and the mid-1980s because of daring and successful product designs, beginning with the TF39 and ending with the F110 and F404. In part, it was also the result of commercial daring: offering fixed-price multiyear contracts to the U.S. DOD, which had been smarting from the overruns of its previous development contracts with Pratt & Whitney. And, in part, it was the result of stringent attention to product and life-cycle cost in design engineering and production. The growth in market share came by plan, not by accident.

As a consequence of these competitions, the U.S. Air Force and U.S. Navy got better engines from both Pratt & Whitney and GE at lower prices, a benefit for the U.S. military, the American taxpayer, and of course for GE as its market share grew. The commonality of design concepts and sometimes of core engines introduced by Neumann and his advanced product planning operation allowed military and commercial engineering programs to support each other technically, while the production of common components drove manufacturing costs down even further; the complementary streams of income reinforced GE's ability to fund advanced product development.

Chapter 9

Competition as a Factor in the Military Market

CONVENTIONAL wisdom maintains that there is almost no competition in the military market after the government customer has selected a product design. Governments generally protect their producers of military equipment and are believed to close their internal market to foreign suppliers. Buyer and seller argue about *allowable costs*, not about *prices*: the higher the cost, the higher the profit, even if it is only a low percentage of the selling price. Even when the fee is "fixed" and does not grow in case of cost overruns, it is negotiated before work begins as an amount acceptable *in proportion* to the total contract cost. Thus, the higher the estimated cost of the work to be done, the higher the fee will be. The negotiated profit margin is usually low because large profit margins on purchases by a government are not politically acceptable. In the United States, interest on loans for work in process is not an allowable cost to the government. That combined with low profit margins does not encourage the manufacturer to take large financial risks. Therefore, typically the manufacturer receives progress payments so that it does not need to finance its own work in process. The commercial market is different, of course. Selling price is set by competition, and costs are of concern only to the seller, who must make sure that they are lower than his selling price.

As the threat of war between the former Soviet Union and NATO receded, and as governments began to face the escalating costs of developing high-performance military aircraft, the military engine market began to take on more of the economic aspects of the commercial market, particularly in the effect of competition on price. As early as 1968, 10 years after the introduction of the F-104 Starfighter and facing one delay after another in the delivery of its new Tornado, the German air force selected the McDonnell F-4 Phantom as an "interim" fighter and ordered 175 of them. MTU, which had assembled some 800 J79 engines for the Starfighter program, expected to build the more than 400 J79 engines needed for the Phantoms, under an extension of its J79 license agreement with GE, but, to its surprise, GE competed aggressively for the contract—Jack Parker himself gave the order to go vigorously after the German business. At the time, GE was delivering large numbers of engines to the U.S. government for the war in Vietnam, and its cost for the J79 had come down with high production rates; MTU was unable to match GE's costs. Nevertheless, the German government awarded a contract to MTU, but at a price determined by GE's offer, far lower than MTU had bid originally. To meet the price and cut its costs, MTU had to order about 40% of the engine from GE while it produced the rest in Germany. Some of the MTU managers grumbled that they could have negotiated the same production-sharing with GE, but at the higher original MTU price, which would have been more profitable for both companies. As classical economics predicts, competition had worked to lower the price to the benefit of the consumer.

Two Italian programs illustrate the effect of absence of competition. In the early 1970s, the Italian government decided to develop the AMX attack aircraft in partnership with Brazil, using the Rolls–Royce Spey 801 engine, a version of the Spey used in the British Royal Air Force Buccaneer attack aircraft. There was a political dimension to selection of the Spey: Italy was determined not to use an American engine because, when two years before FIAT received an order from Libya for 20 G.222 transports, the U.S. government would not approve the transfer to Libya of the T64 turboprop engines built by FIAT under GE license. At the time, Colonel Ghadaffi's government owned 11% of FIAT's common stock. FIAT replaced the engines with Rolls–Royce Tyne engines, at great expense and with reduced altitude performance, and delivered the aircraft to Libya. The Italian industry vowed that in the future it would avoid the use of American equipment where possible, to avoid interference by the United States in other export opportunities, such as when Idi Amin of Uganda briefly expressed interest in the G.222. Because Italy hoped to export the AMX, it needed to find an engine outside of the United States.

The design of the Rolls–Royce Spey used in the AMX goes back to the early 1960s. For the AMX, it is assembled by FIAT with parts from Rolls–Royce and Italian manufacturers. The selling price to the Italian government is about $3 million per engine, roughly the same as the price of a GE F110, the much more modern and powerful engine for the F-16. The high price for the Spey may be caused by low production rates, but it may also reflect the fact that the Italian airframe industry and the government did not seriously consider any other engine for the AMX program.

Governments tend to define the manufacture of certain products as critical to national prestige and survival, but there is a price for self-sufficiency: they must pay whatever it costs to design and manufacture such products in small quantities. When international economic barriers are lifted, previously protected national manufacturers usually cannot compete based on cost against larger international manufacturers. A good illustration of this is what happened after Franco died and Spain was finally able to join the European Community, and Spanish manufacturers who had enjoyed a protected national market suddenly found themselves unable to compete against larger European manufacturers or against imports from the United States. A similar economic mechanism is at play among the airlines of Europe: small national carriers that thrived under bilateral cartel agreements to share traffic and revenue found themselves unable to attract passengers or compete based on cost with larger, more efficient carriers, once bilateral limitations give way to competition for European traffic.

The Italian aviation industry is noted for its creative inventors and manufacturing craftsmen. Mussolini fostered the development and manufacture of military aircraft and engines, whose performance was limited more by chronic shortages of capital and high-grade materials than by any lack of good ideas. At the end of the war, the industry had only one customer, the Italian government. Like most national industries, it was far bigger than its one customer needed or could afford. By a regulation dating back to the Mussolini era and never rescinded, the Italian defense ministry can pay contractors only the hourly wage rate then prevailing. Because the ministry is well aware of actual costs 50 years later, it connives at an artificial accounting that accepts inflated labor hours multiplied by the outdated wage rate. Ever since 1945, the industry has danced around the fact that the Italian market is too small to support its present size, unwilling to consolidate or reduce capacity. To survive, the industry and the government have avoided competition, steered programs to each company in turn, and allocated shares to the others, and the government has paid the premium involved in keeping the industry larger than its market justifies.

This is not a purely Italian phenomenon, of course. In the early 1980s, the Swedish parliament debated what fighter to buy to replace the Viggen, which had been designed and built by SAAB. The Swedish defense ministry offered three alternatives. The cheapest was to buy fighters complete from the United States, but Sweden would have to add to the purchase price the intangible cost of giving up its own capability to design fighters, and the tangible cost of shutting down its aircraft and engine factories. The next option was to build an American fighter under license, which would have cost perhaps 50% more than importing and would have preserved manufacturing capability, but design capability would still have been lost. The most expensive option was to design and build a Swedish fighter with an American engine and avionics built under license (designing all three systems was beyond Sweden's means), the only option that would preserve Swedish design capability and its capability for neutrality. Any of the options would have satisfied the requirements of the Swedish air force; the choice between them was purely political. Parliament chose the third option, the most expensive.

The Italian engine industry is composed of three firms: FIAT, Alfa Romeo, and Piaggio. FIAT is a large automobile manufacturer whose stock is controlled by the Agnelli family. FIAT's airplane division was spun off in the 1960s to the state-owned firm Aeritalia, which collected most of the wartime airframe manufacturers under its corporate roof. FIAT's engine division has been the prime contractor for the Italian government for large military engines and has also specialized in the design and manufacture of precision gears for engines and helicopters. In an attempt to diversify from military engine production, FIAT entered into a joint venture with GE in 1968 to develop marine and industrial versions of the TF39 turbofan. The very successful LM2500, a 25,000-hp gas turbine, came out of this collaboration. All of the U.S. Navy's destroyers and frigates use the LM2500, as do many other navies. The engine has also found application on oil production platforms, in pipeline pumping, and in electrical power generation. Following the successful LM2500 experience, FIAT joined GE in the CF6-80 program, designing low-pressure turbine blades, manufacturing parts, and doing development engine testing. FIAT later took a small part of the Pratt & Whitney PW4000 program, and finally joined the International Aero Engines syndicate to develop and build the V2500 turbofan.

FIAT succeeded brilliantly in the technical aspects of these collaborative programs, but the commercial side continues to present great problems. Italian aerospace hourly wages and benefits are higher than those in the United States, while American productivity tends to be higher. FIAT grew up in an atmosphere of having the Italian defense ministry pay for whatever a military program cost, but the LM2500 and the commercial engine programs that followed were governed by different rules: competition in the market set the price, and costs had to be less than that to make a profit.

The problem was equally acute at Alfa Romeo Avio, the aircraft engine division of the state-owned automobile manufacturer Alfa Romeo, which was later absorbed by FIAT. After the war, Alfa Romeo Avio had been the Italian government's prime contractor for small aircraft engines, primarily for helicopters. Alfa Romeo had designed a 500-hp turboprop of its own, but it was neither better nor cheaper than others from Lycoming, Allison, or Turbomeca, and it came to the market several years later. Needless to say, the Alfa Romeo engine did not go into production. Alfa Romeo also had a significant engine overhaul business for the Italian air force and for the national airline Alitalia, which was owned by the same state holding company that controlled Alfa Romeo Avio and the airframe firm Aeritalia. When Alfa Romeo decided to diversify into commercial engine programs, it had great difficulties accommodating itself to the tyranny exerted by market price over manufacturing costs. Alfa Romeo's first venture was the GE CF6-32, the engine for the Boeing 757. After Alfa Romeo and the other partners had all made their investments, GE terminated the program because it was not competitive against the superior Pratt & Whitney PW2037, nor the Rolls–Royce RB.211-535, whose price was presumably made competitive by the British government.

After that dreadful experience, Alfa Romeo became very cautious about joining another program. Only after several years was its managing director able to convince his board that Alfa Romeo had to participate in international commercial programs to survive at all. Alfa Romeo became a major manufacturing partner in the Rolls–Royce Tay program, and has taken a small but important role in the CFM56 program with SNECMA and GE. There have been discussions for more than 30 years about a merger between the FIAT and Alfa Romeo aircraft engine companies, and it is possible that such a merger will come about as the European Union eliminates economic barriers and countercompetitive practices. The unknown question is whether such a merger would also accomplish the reduction in capacity dictated by the market.

The third company, Piaggio, is part of a family-owned conglomerate that also manufactures aircraft and motor scooters. The engine division has been the prime contractor for the Italian army's helicopter engines, able to get lucrative contracts for the manufacture and overhaul of small numbers of engines designed by Lycoming and later by Rolls–Royce. When the Italian helicopter firm Costruzzione Aeronautiche G. Agusta designed the A129 twin-engine helicopter and sold it to the Italian army and other military services, it selected the Rolls–Royce Gem 850-hp turboshaft engine, which Piaggio assembles from parts made by the Italian industry and Rolls–Royce. In the early 1990s, the Italian Army paid some $750,000 for each Gem, about the same as it paid Alfa Romeo for CT7 turboprop engines of twice as much power and with lower specific fuel consumption.

In 1983, the defense ministries of Britain, Germany, and Italy began planning for a new heavy helicopter as a transport for their armies and air–sea rescue and for maritime patrol by their navies.

The project was given the designation EH101 (EH stood for the symbolic "European Helicopter"), with Westland Helicopters in Britain and Agusta in Italy as the design partners. The EH101 would be equipped with three engines as an additional safety factor in its maritime role. Britain proposed the RTM 322 engine, to be developed jointly by Rolls–Royce and the French engine company Turbomeca, and perhaps by MTU in Germany. Since the early 1950s, Rolls–Royce's helicopter engine factory in Leavesden built T58 turboshaft engines under GE license. The original agreement had been signed between GE and Bristol–Siddeley Engines, Ltd., which was subsequently merged into Rolls–Royce. Bristol–Siddeley had also started the development of the Gem engine. In the late 1960s, a Rolls–Royce executive, the retired RAF air marshall Sir Denis Spotswood said that "helicopter engines were too big a market to be left to the Americans." France did not select the EH101, but elected to go its own way with incremental improvements of an existing French helicopter, the Aerospatiale Super Frelon, which could also use the RTM322.

The biggest fleet of military helicopters in Europe is that in Italy. The three armed services and the Carabinieri, the national police, use about 300 medium helicopters of the "Hughey" class made familiar by the war in Vietnam, in addition to 30–50 heavy helicopters and more than 50 light helicopters. Agusta built Hugheys, Sikorsky medium and Boeing heavy helicopters under license, and designed its own A.109 and A.129 light helicopters. The Italian government recognized that it would eventually need to replace its Hugheys with more modern machines, and GE and Alfa Romeo proposed to develop a single engine type to be used in both the Hughey replacement and the EH101, based on the successful T700 engine GE had developed for the U.S. Army. Standardizing one engine type made economic and logistic sense, and the T700 had already demonstrated outstanding reliability and fuel economy in the U.S. armed services.

The Italian government therefore funded Alfa Romeo for the development of the new engine with the civil type designation CT7-6, in cooperation with GE; having done so, there was no doubt that it would select this engine for its EH101 helicopters. By the same token, the French, who had no intention of buying the EH101 for their own armed forces, proposed that the Turbomeca/Rolls–Royce RTM322 become the NATO standard engine in its class; such a standard might be construed to apply also to Italy, a more direct NATO member than France. The third party, the British Ministry of Defence, decided to hold an open competition for the engines for its EH101 helicopters, inviting GE to bid the CT7 against the Rolls–Royce RTM322 and holding out the prospect of an additional market in Hughes Apache antitank helicopters, which the RAF expected to order some years in the future. In the U.S. Army, the Apache uses the GE T700 engine; in the RAF, it could use whichever engine type was selected for the EH101.

Many GE people were convinced that Rolls–Royce had an unfair advantage. In their view, Rolls and Turbomeca had a paper engine and could promise anything; GE had an existing product with a proven track record but also with an upper limit on its possible power rating. Nevertheless, GE competed vigorously. With Alfa Romeo, GE laid out a growth program to increase the power available from the CT7-6 during emergencies, to the level GE suspected the British and French had set arbitrarily high with the intention of eliminating the CT7-6 from the competition. GE and its Italian partner responded to every step of the competition with what they considered a winning technical proposal, and with a very challenging price. Nevertheless, Rolls–Royce won the competition. The two proposals were judged more or less equivalent technically, with more growth potential in the RTM322 and a proven record of reliability in the CT7. But Roll–Royce's price was lower, to GE's great surprise.

Although GE grumbled about the result, it could be argued that it was a small victory for GE and for the principle of competition. It was the first time that the British Ministry of Defence held such an open competition. To win, Rolls–Royce had been forced to underbid the GE/Alfa Romeo team, to the benefit of the Ministry. And Rolls–Royce lost the opportunity to earn monopoly profits on yet another military program such as the Gem in Italy, from which it could have fed funds to its commercial engine developments.

Is it realistically possible for foreign engine firms to succeed in markets preserved to the national engine company, such as those in

Britain and France? That depends on the degree of government ownership of the customer, on the size and political importance of the order, and whether a suitable local product is available. Thus the French Navy was allowed to buy eight Vought F-8 Corsair II fighters from the United States for one of its early aircraft carriers but was later forbidden to buy F-18s because Dassault was developing a naval version of the Rafale. The French Navy's few carrier-borne radar planes were bought from Grumman—there is no practical European equivalent—but French military helicopters all come from Aerospatiale, their engines from Turbomeca. In contrast, Air France, citing its purchase of some airliners from Boeing and others from Airbus Industrie, maintains that it can select whatever equipment it needs, whether there is French content or not, despite the fact that it is owned by the French government. However, if there were head-to-head competition and performance were truly comparable, there is little doubt that Airbus Industrie would be selected over Boeing. It must be noted that SNECMA makes no commercial engines of its own other than the CFM56 in partnership with GE. Air France, therefore, has no problem buying American or British engines for its aircraft, but the French government no doubt smiles when such an engine has a healthy SNECMA content. For example, SNECMA contributes more than 30% to the GE90 engine in the Boeing 777.

In Britain, the RAF first bought FB-111 bombers, and later F-4 Phantom fighters with the engine changed to the Rolls–Royce Spey, from the United States after Duncan Sandys cancelled development of the TSR2 long-range bomber. There was such a political outcry about the damage these imports caused the British aerospace industry and the British economy as a whole that subsequent British governments acquiesced to cooperative European development programs such as the Tornado (as an attack aircraft and in an interceptor version) and the Eurofighter, despite their high cost. A British program was funded for development of a radar for maritime patrol and airborne early warning, to be carried on some of the original Comet IV airframes. Only when the costs exceeded target by several orders of magnitude and Marconi, the prime radar contractor, had to concede that the radar still did not perform to specification, did the British government give approval for the purchase of Boeing E-3B Airborne Early Warning and Control Systems (AWACS). The E-3 is a 707 carrying a Westinghouse radar, whose large antenna is mounted above the fuselage in a rotating aerodynamic fairing. The cabin is full of transmitters, receivers, and control consoles.

On the civil side, British Airways bought British airliners when it was government owned. Absent a suitable British aircraft, the airline bought Boeing 747s for its long-range routes, and Rolls–Royce made sure that there would always be a Rolls–Royce engine available, even on airliners of American manufacture. However, since British Airways has been privatized it has not hesitated to buy GE engines when the commercial deal was attractive enough, for example, for its Boeing 777.

Clearly, competitive forces can play a role in military programs and will have the restraining effect on price predicted by classical economic theory, even though sometimes competition is distorted in favor of a national champion. When competition is limited or excluded by political decision, monopoly prices result, which makes it surprising that governments are so reluctant to take advantage of competition. There are obviously occasions when the lowest price is not the most important national objective.

CHAPTER **10**

■

License Agreements and Offset Programs

IN THEORY, at least, commercial aircraft and their engines are bought strictly on the basis of performance and competitive price, with no consideration given to local production or industrial policy. In practice, countries where the airlines and the aircraft and engine industries are government owned have always been suspected of favoring their own manufacturers when it comes to buying airliners. Some emerging economies without an aircraft industry, for example, Australia, Singapore, and the Republics of Korea and China, have regarded equipment purchases by their airlines as opportunities to press for industrial benefits of one kind or another, if only through an infusion of modern technology. But most airliner purchases are too small to justify local production. The lead time would be too long, the investment in tooling and facilities too high, and the unit costs not competitive against those of the original designer's factory. Besides, airliners are intended to generate a stream of income from ticket sales, which presumably recovers the investment.

The purchase of foreign military aircraft raises different political issues, particularly in countries with aircraft- and engine-manufacturing industries of their own. When a foreign military aircraft is imported, money flows out of the country, but there are no ticket sales to recover the investment. If the need is for a small number of special aircraft, for example, Vought F-8 fighters for a French aircraft carrier, the sums may be small enough that there is no pressure to develop a new indigenous system, although it has become conventional for the buyer to ask the seller to make offsetting purchases. Sometimes the foreign product is so unique that it makes no sense to consider duplicating it, in which case the foreign product is acceptable provided that the seller offers substantial "offset" purchases to his local competitor. The Boeing E-3 AWACS falls into this class. Britain spent millions on the development of the Nimrod maritime patrol aircraft and its Marconi radar, without bringing the program to a successful conclusion and, after much public embarrassment, ordered the Boeing AWACS as France had done earlier. Boeing and its subcontractors CFMI and Westinghouse had to commit to off-setting purchases from the French and British aerospace industries.

The situation is no different in the United States. When a foreign aircraft or engine is uniquely suited to an American mission—for example, the British Aerospace T-45 trainer or the AV-8B Harrier VTOL fighter for the U.S. Marine Corps—the United States will buy the foreign equipment but insists on substantial production in the United States. The stated reason is always self-sufficiency in case of war, but no doubt industrial policy plays a role: protecting the local industry can be rationalized as a factor in military self-sufficiency. Democratic governments are to one degree or another politically accountable for how they spend public money—therefore, the emphasis on audits and control of military program costs; the industrial spin-off from military technology; and the creation of jobs from government procurement. Whenever a foreign system is to be adopted, there is a logical progression in military industrial policy. If the production program is big enough and the indigenous industry is a large source of employment, governments often insist on local production under some form of license from the foreign system designer.

During the Cold War, the United States had the biggest military establishment outside the former Soviet Union, and invested more funds in the development of advanced weapons than any of its allies. There were many American military aircraft models to choose from, and they were often more capable and less expensive than those developed by other countries. For the sake of mutual security, the United States was relatively generous in making technology available to its allies under license. Naturally the flow of technology was preponderantly from the United States to the allies. That created some resentment, particularly among the technical fraternity in the European countries, which believed that its own military industry was being driven to irrelevance by competition from American weapon systems. Even production under license was perceived as a conspiracy by the United States to hollow out what the Europeans called "the noble work" of design and development.

Under such license agreements, the buyer gains the right to make, use, and sell the seller's proprietary system, using the seller's data and knowledge, while the seller agrees not to assert any patents it may have against the licensee. The license agreement usually specifies geographic limitations, particularly territories where the licensee may sell products made under license. And because governments generally regulate the sale of military equipment, the approval of the seller's and the buyer's governments is usually also required for export sales to third-party countries. Government approval can involve issues of foreign policy, for example, when the Italian government decided not to approve the sale of fighter aircraft to Taiwan to protect its interests in the People's Republic of China, or when the Spanish government bought AV-8B fighters for its aircraft carrier and insisted that they come from U.S. production, not from Britain. The issue then was Britain's "intransigence" over Gibraltar. There may be security issues—is it likely that licensed systems will be used against the licensor's own government? For example, the Aeritalia G.222 transport uses a GE engine built under license by FIAT. If Aeritalia were to sell such an aircraft to Libya, does Libya pose a potential military threat to Italy? Would such a sale be in the interest of the U.S. government?

And finally, there is the economic issue—is it in the interest of the seller's government to create potential competitors in the form of licensees or to hollow out its own military industry by exporting subcontract work to another country? Different governments give different answers. France and Britain have always put a lot of emphasis on preserving employment in their home aerospace industries. U.S. Presidents Reagan's and Bush's administrations denied that the United States practiced such "industrial policy." It was their ideological faith that the aggregate of decisions made by individual companies, in other words, by "the market," would always be more efficient than and superior to decisions made by government officials, and they pointed to the economic collapse of the USSR as convincing proof of the dogma. Thus, the Bush administration dismissed the director of the Defense Advanced Research Projects Agency for the ideologically incorrect act of funding the development of semiconductor manufacturing technology so that the United States would not have to depend on imports from Japan and South Korea.

In spite of these political claims, the U.S. DOD protects what it calls the U.S. defense industry base and, in the case of fighter engines, carefully divided orders between Pratt & Whitney and GE so that each firm got precisely half of the orders. The U.S. Government approved GE's and Pratt & Whitney's European and Japanese license agreements for fighter engines because it is clearly in the overall U.S. interest that allied countries equip their air forces with fighters compatible with those in U.S. service, at their expense rather than with U.S. funds. Such agreements also lead to substantial export sales from the United States to supplement what licensees do in their countries. Even when the U.S. DOD interest is purely military, for example, when it limits the export of high-temperature turbine technology to ensure that the tactical performance of American engines remains in the lead, or that their characteristics as targets are kept secret, other governments suspect that the United States is using this as an excuse to maintain U.S. commercial hegemony of the arms export market—just as they would do if they could.

Despite these conflicting objectives there have been practically no deliberate violations of the terms of such license agreements. Some small firms in Taiwan have made spare parts for J85 engines without the benefit of a license agreement, using data obtained by the Republic of China government from the U.S. Air Force; some of these subcontractors have offered their parts on the world black market for "bogus" parts. Reputable users avoid the use of such bogus parts because they raise questions of liability in case of an engine failure. In general, licensees and their governments have depended too much on the U.S. government and the licensor's technical support to risk any small gains they may make by acting outside the limits of an agreement.

If a military production program is not big enough to justify the tooling investment and learning costs of local production under license, the buying government may insist instead on offsetting purchases from its industry by the system seller. That balances payments in national accounts, and creates some work for the local industry. The Canadian government was the first to insist on offset purchases as a condition of its adoption of the GE J85 engine for a Canadair trainer and, later, of a more powerful version for the Northrop F-5 fighter (Canada and later the Australian government refined the accounting for offset to a major bureaucratic art). Offsets then became a common condition of a sale in countries such as Britain and France that wanted to protect their aerospace industries, and in emerging economies that did not want to be left behind and saw an opportunity to create an aerospace industry of their own with the assistance of foreign manufacturers.

Direct offset to a country involves the local manufacture of components for incorporation into the systems being bought by that country, and also for systems made by the seller for use elsewhere. Firms in the buyer's country often have great difficulty being competitive in price for such offset purchases against the seller's existing subcontractors. The American aerospace industry is very competitive in cost, both because of its larger production runs that lead to lower costs, and because of high productivity. Buying governments quickly accept the logic that their own firms have to be competitive to qualify for offset purchases—although in some cases, the government may assist the local subcontractors by paying for tooling or training and, in rare cases, by paying a premium price for the components used locally.

The Turkish F-16 program is a classic example of direct offset and licensed production. When the Turkish air force ordered 160 F-16 fighters from the U.S. government in 1984, General Dynamics agreed to set up a factory near Ankara to assemble the aircraft and make parts for them there. Pratt & Whitney agreed to do the same for the F100 engine. The American companies agreed to be part owners of the factories so that the Turkish Air Force Foundation, the majority owner, would have some assurance of whole-hearted technical support—"chaining the bridegroom's leg to the bed," joked a Turkish general. GE had competed in Turkey with the F404 engine in support of the McDonnell–Douglas F-18, which lost the competition to the F-16. When soon after that the Israeli air force and the U.S. Air Force both selected the GE F110 engine for their F-16s, the Turkish air force asked GE whether it would accept the offset provisions of the Pratt & Whitney proposal, including minority ownership of an engine factory. GE agreed, and Turkey switched to the GE F110.

The Turkish air force operates hundreds of GE engines in its F-104, F-5, and F-4 fighters, and many GE field service engineers work in Turkey to support them. In Eskisehir, in Anatolia, GE had previously helped the Turkish air force set up and run an overhaul depot for its engines. The new engine factory was built next to the air force overhaul depot, on which it drew for its board of directors and operating managers. The work force came from among the thousands of Turkish who had been trained as "guest workers" in Germany, many of them in the aircraft engine factories of MTU and Klockner–Humboldt–Deutz (KHD). GE trained the workers, technicians, and managers and provided the work load to the factory for F110 engine parts for GE's worldwide markets, as well as making parts there for the Turkish program and assembling the engines. Because of the experience and skill of the Turkish work force, its relatively low wage and benefit costs, and the level of work occasioned by the Turkish F-16 program, the Eskisehir factory was quickly able to attain competitive cost levels. But as is the case in any such offset program, once the Turkish F-16 program was completed it was difficult to find other work for the factory. There is already enough capacity in GE's own factories to handle worldwide demand for F110 engines.

In many cases, it is impossible for local firms to be competitive in price. In pure licensed production, the government may elect to pay whatever it costs the local firm to build the engine. But when offset is required and there is no qualified and competitive local manufacturer, offset must become indirect. The selling firm may buy non-aerospace local products for export or facilitate such purchases by third parties. In Turkey, GE arranged for the export of steel reinforcing bars, phosphate minerals, polyester sacks, and other commodities. This is obviously less attractive to the buying government, which usually believes that it can export goods and services in which it has a comparative advantage without assistance from the military system seller. The seller may be asked to transfer other manufacturing technology or supply other information, as a form of offset. There is a two-stage negotiation here, first between the system seller and the buying government on the amount and form of offset. Once they have agreed, however reluctantly, to an offset deal that is realistically attainable, they join forces to convince the political public in both countries that the agreement is good for both countries—not merely for the selling firm and its customer alone. In the case of highly specialized products such as aircraft engine parts, local small-scale production makes little competitive sense. As soon as the immediate offset obligation has been satisfied, the system seller has no more incentive to put work into a local manufacturer or to accept other than competitive prices and often also faces a demand for offset from his next customer. In the worst case, the buying government has invested in a manufacturing facility for whose output there is no demand. The sharpest example of this is the former Soviet aerospace industry. With production capacity great enough to satisfy demand at the peak of the Cold War, today its vast factories stand largely idle. In the United States, consolidations of the industry and the closing of factories have matched capacity to the current level of demand, at the price of forcing out thousands of skilled workers, engineers, and managers to find other work on their own.

This has led governments to encourage other forms of offset, for example, the establishment of companies to repair and overhaul aircraft engines, where the capital investment, the primary barrier to entry, is much lower than for the manufacture of engine parts. Such overhaul depots are intended to maintain the imported aircraft and engines, and to offer maintenance and overhaul services to customers in other countries. GE helped to set up such an overhaul depot in Greece, Hellenic Aerospace Industries; in Indonesia, the Universal Maintenance Center; and in China. There is a ready market for such services when an overhaul depot is the first in a region, but because the barriers to entry are low, others can enter the market and very quickly create an excess of capacity. As a result, there is fierce price competition. Because it is feasible to transport engines by air for overhaul, the competition can be worldwide and include the most productive competitors. At one time, for example, the government of Saudi Arabia wanted to set up such an overhaul depot as an offset to large defense purchases from the United States, which had themselves, ironically, been made to offset a huge increase in petroleum imports from Saudi Arabia resulting from the Organization of Petroleum Exporting Countries (OPEC) cartel's price increases for crude oil. GE proposed the Middle East Propulsion Center, an engine overhaul depot modeled on Hellenic Aerospace Industries. GE would own a minority of the shares, and would furnish information and train Saudi technicians and managers. The work load would come from the engines in the Royal Saudi Air Force and Saudia, the national airline. But the Saudi Ministry of Defense & Civil Aviation insisted that the price of overhauls performed by the Middle East Propulsion Center would have to be competitive, difficult when Saudia has the option of sending its engines to Lufthansa or to the United States for maintenance. GE proposed that the Saudi government enlarge the work load by calling on countries receiving Saudi financial aid, such as Pakistan, to send their engines to the Middle East Propulsion Center for overhaul. Saudi Arabia is traditionally reluctant to put pressure on its neighbors. Particularly where little prior local industrial capability exists, it takes a lot of time and patient negotiation to reconcile the desire for creating local work of high added value with the requirement of competitive costs. Negotiations about the Middle East Propulsion Center went on for years. Construction and the hiring and training of staff took more time. It is not clear whether the operation will ever be successful other than as a token of the determination of both parties to find a solution.

Sometimes prestige may be more important than economics. In the mid-1970s, the Shah of Iran arranged for a German company to set up the manufacture of truck diesel engines in Tabriz. The Germans erected an automated transfer machining line, imported from Germany. Swarms of German mechanics adjusted the machines and kept them in good operation. At one end of the line, four Irani workers loaded rough iron castings onto holding fixtures. At the other end of the line, four more Iranis unloaded the finish-machined engine blocks. Obviously the value added in Iran was low.

When the Greek government decided to order F-16 fighters, perhaps because the Turkish had already done so, realists in the government rejected the notion of more offset. Hellenic Aerospace Industries was in operation, technically qualified but with a limited market. (It had won an overhaul contract from the U.S. Air Force and Europe against strong competition from Britain, Israel, and the United States). Instead, the government accepted a bold proposal from three American contractors, General Dynamics, Westinghouse, and GE, who set up and managed a venture fund and subscribed its capital of $50 million. The objective of Hellenic Business Development & Investment Corporation (HBDIC) was to provide venture capital and management advice to Greek start-up operations that had the potential of creating exports. The HBDIC would take an equity in such companies, with banks and local investors providing the rest of their working capital. As such companies prospered, HBDIC would sell the shares and make the income available for further investments. Foreign ownership and management of HBDIC was intended to keep the $50 million capital free of political diversions and make available impartial advice to Greek entrepreneurs on such basic matters as business plans and market research, thwarting the Greek penchant for political interference in the economy and for siphoning money from private companies to political parties.

The HBDIC considered a wide range of investment opportunities: a large tuna-fishing boat with a contract to sell fish to Mitsubishi Trading Company; a marble quarry; a resort center; a chain of health clinics offering a high standard of diagnosis and treatment to rich clients from all over the Middle East; and a factory to import circuit boards from the United States for assembly into personal computers to be exported to Russia. Some of these schemes turned out to be fanciful, but they rose or fell on the realism of their market analysis and business plans, a standard no direct offset could have met. The $50 million investment (perhaps the price of one or two F-16s?) freed the three American companies from the continuous agony of trying to set up Greek aerospace manufacturers capable of competing on cost, and most likely disappointing them and the Greek government when that proved impossible.

In most of these cases, military imports were the immediate justification for demanding offsets. Occasionally, governments have attempted to influence an airline's selection of competing aircraft by granting landing rights to foreign airlines or threatening to withhold them. Some governments have asked for offset purchases when their national carriers buy foreign airliners; some airlines themselves have asked for offset purchases simply to get money to pay for the import, for example, JAT in the former Yugoslavia; sometimes a private airline is owned by a conglomerate that seeks offset benefits for another part of the enterprise.

Importing countries see such offsets as similar to import duties, protection against powerful market leaders, while exporters tend to see them as economic distortions of normal trade. The United States, the world's biggest aviation exporter, is of course opposed to such distortions and led international negotiations of the Civil Aircraft Annex to the General Agreement on Tariffs & Trade. Signatory governments have agreed not to demand offset for airliner purchases nor to involve extraneous issues such as landing rights. The governments have also agreed not to subsidize the financing of airliner sales into each other's countries. Airline equipment is to be selected strictly on the basis of performance, price, and delivery, although sellers remain free to offer offset if they consider it in their own competitive interest.

In the heat of a sales competition, when the customer asks for offsets, technology transfer, or a license agreement as a concession, the marketing people are often quick to agree. They immediately understand the significance of a price concession or a guaranteed maximum cost of maintenance per flight-hour. But the costs of technology transfer or of agreements to purchase from a new vendor in another country are less tangible, and perhaps they get lost on the books of the manufacturing division and never show up in the direct profit-and-loss calculations of the sale. Nevertheless, offsets are a legitimate part of a marketing strategy. During the Canadian competition for the New Fighter Aircraft, the Canadian government announced that it would judge competing proposals not only on performance and price, but also on the amount and quality of offset created in Canada. GE's F404 engine was in the F-18, the aircraft ultimately selected, and GE proposed setting up a factory in Canada to manufacture compressor blades for its commercial engine production in the United States. Did GE agree to transfer work from its American factories or suppliers to a new Canadian facility? GE says no. GE makes all of its own compressor blades, until then at a factory in Vermont that was already running at peak output; GE needed to add manufacturing capacity. It could not do so in Vermont, where the blade factory and another making aircraft cannon for the U.S. DOD already drew on much of the available labor pool, but it could build a factory elsewhere in the United States—or in Quebec, right across the border from the Vermont factory. Building the Quebec factory was a factor in the F-18's winning the New Fighter Aircraft competition. GE maintains that building it and winning the Canadian competition resulted in more exports of F404 engines and parts to Canada than imports of blades from Quebec.

A company like GE, with its broad product lines, worldwide markets, and global production, has a theoretical advantage in juggling such offset transactions, but historically there has been no inclination among the various GE businesses to sacrifice their independence to benefit another GE business. And even in countries where a global company such as GE has a big presence, the next transaction usually creates its own competitive pressure for offset. GE's important corporate presence in Canada gave its offset proposals great credibility, for example, during the competition for the EH101 helicopter engine some years after that for the New Fighter Aircraft, but GE's proposals were always weighed dollar for dollar against the competitor's proposals. What is past is done, in Canada and elsewhere.

Chapter 11

The Foreign Corrupt Practices Act

After Watergate, a U.S. Senate committee headed by Senator Frank Church of Idaho went on the prowl for illegal corporate contributions to the campaign to reelect President Nixon. The committee found that large sums of cash had been sent by the chairman of Northrop to Albert Savy, a Northrop consultant in Paris, supposedly as his fee for doing work for Northrop. From him the money went through intermediaries in Mexico and ultimately to Herb Kalmbach, the treasurer of Nixon's campaign. In the course of pursuing its political investigation, the committee learned that such large payments, again some in cash, appeared to be not unusual in the course of marketing military aircraft to foreign buyers. Northrop defended itself by claiming that it was doing no more than it had learned from Lockheed in Europe and in the Middle East. The committee quickly questioned Lockheed and heard allegations that large sums had been paid to prominent officials, among them the late German defense minister Franz-Josef Strauss, the late Prince Bernhard of the Netherlands, and several Saudi generals, all to secure orders for F-104 fighters and C-130 transports. It was a wonderful brew of big names, big money, and foreign intrigue.

Little of this had to do with campaign contributions to President Nixon, but the sensational nature of the allegations led to the passage of the Foreign Corrupt Practices Act, signed into law in 1977. The Foreign Corrupt Practices Act makes it illegal for an American company to offer bribes to a foreign government official as an inducement to buy a product. It has always been against U.S. law to bribe an American government official, as a matter of public policy, but until then bribing a foreign official had not been considered a crime under U.S. law as long as the bribe was paid out of profits and was not deducted from revenues before corporate taxes were paid. Similar rules applied then to foreign companies, and in most countries continue to do so to this day.

A company such as GE does business in many countries and is therefore particularly vulnerable to actions that might violate the Foreign Corrupt Practices Act. Any infraction would have a serious impact on its core business in the United States and could result in the withholding of export licenses by the U.S. government. GE, therefore, put strict rules into effect to make sure that its divisions neither offered nor paid bribes, and did everything a prudent person could be expected to do to make sure that no third party would do so on GE's behalf.

Why use such third persons at all? For the same reason that General Motors sells cars through independent dealers or that other firms use manufacturers' representatives and trading companies: no firm can afford to have its own employees everywhere. Cost, language, culture, law, and an understanding of the local business community lead a company to serve certain markets through sales representatives, who solicit orders on its behalf. In fact, the laws of some countries, notably those of the Arabian peninsula, mandate the use of local citizens as sales representatives for foreign companies. Other markets are best served through sales distributors who buy products from the manufacturer at a discount from the list price and resell them to customers at the list price, often carrying inventories of products and spare parts, and offering repair and overhaul with mechanics trained in the seller's factory. Sometimes it is important to have available the services of a consultant who can advise the factory on the objectives of the customer and can identify who will make the decisions, as well as sounding out the prospects for a sale and making discreet introductions. There is nothing unusual or illegal about such methods of distribution: they are common in the United States and Europe as well as in the Middle East, for automobiles and shoes and groceries. Nor is there anything unusual or illegal about how such commercial "agents" are paid: distributors buy at a discount and may resell at the manufacturer's list price or at a higher or lower one; sales representatives receive a commission when a successful sale has been made, and sometimes also have their market development expenses reimbursed by the factory; consultants usually receive a retainer and expenses, whether a sale results or not. Sometimes they receive a bonus if a sale is made.

It is the paying of bribes that is illegal under U.S. law. Proving that a bribe has been paid will never be easy, and clever criminals find ingenious ways to launder the transaction. The payment of unusually large fees or commissions always raises the presumption that some of the money has been used for illegal purposes and necessarily attracts attention. But how big is a large sum of money? A joking response in the 1970s was, "Any sum greater than the salary of an investigative reporter." A sales commission of 1% of the purchase price does not seem unreasonable for the successful marketing of capital goods, particularly when the sales campaign can last several years. But 1% of a $300 million sale to an airline is $3 million dollars. That is a sum large enough to attract attention.

GE put rules in place to examine prospective agents to make sure that they had an established business presence and were in fact capable of doing what GE wanted them to do, that they had a good business reputation, and that they understood clearly that GE did not want them to pay bribes on GE's behalf. GE also retained the right to terminate any agreement with an agent if he acted in a way that adversely affected GE's business. These rules were emphasized and strictly enforced, even at the occasional expense of effective performance by the agent. GE salesmen used to grumble when they lost a sale that the cause was bribes paid by competitors or their agents. That is usually impossible to prove, and even to the extent that it may be true, it did not prevent GE from becoming the biggest of the aircraft engine companies in revenue. The results of the policy were satisfactory: although it had many such agreements, GE managed to stay largely out of trouble with the Foreign Corrupt Practices Act. Nevertheless, an example will illustrate how difficult that can be.

Britain pulled its military presence out of the Persian Gulf in 1962, and the United States had taken over the task of maintaining political stability in the region. After the U.S. Central Intelligence Agency helped to overthrow the regime of Iran's Prime Minister Mossadeq, who had shown some hostility to foreign oil companies, the United States embraced the Shah of Iran as the focus of power and stability in the region. Saudi Arabia at the time was a minor political factor, although the magnitude of its petroleum resources was already well known.

During the Mossadeq interlude, the Shah of Iran and his family fled to Rome in a military transport airplane piloted by his brother-in-law, who was shot in the shoulder in the course of their escape from the imperial palace. As his reward, General Khatami later was made the commander of the Imperial Iranian Air Force. In Rome, the Shah and his family stayed in a villa owned by an American banker, the late Richard Klehe, who advanced the Shah $1 million so that the imperial family would not want for pocket money during its short exile. After the Shah had been restored to his throne, Klehe naturally remained his trusted friend. Klehe had acted as a financial advisor to GE when GE sold its interests in the computer company Machines Bull to the French government. It was easy for GE to retain him for its business in Iran, where he helped in such transactions as the sale of a big thermal electric power station and the setting up of a factory for producing refrigerators. He remained a valuable adviser to GE until his death in 1969.

In the early 1970s, the Shah began a major expansion of the Imperial Iranian Air Force, buying aircraft with whatever engines they came with. He visualized Iran as the most powerful country in the region and himself as the guarantor of regional peace and stability. That meant that he needed modern armed forces to be able to dominate potential mischief makers, most obviously Iraq. He also wanted to use his armed forces and compulsory military service as a training program for the farmers and shepherds and carpet makers of Iran to bring them into the industrial world. The Nixon administration believed that this policy was in the interests of the United States and assisted the Shah in carrying it out.

Through the U.S. DOD Iran ordered four antiaircraft missile frigates of the Spruance class, the latest U.S. design; the latest U.S. Navy F-14 fighters with advanced radar; large numbers of Northrop F-5 fighters and McDonnell F-4 fighters; Lockheed C-130 transports; and a large fleet of Bell helicopters for the army. Modern tanks and artillery were bought from Germany. The Gendarmerie, a national police force, was organized and run by a former commander of the New Jersey State Police. Many U.S. DOD people, military and civilian, were stationed in Iran as instructors and supervisors to make all of the expensive machinery work. Private companies were brought in to take over some of the training tasks, for example, Northrop and Ross Perot's Electronic Data Systems, which received a contract to set up a system of social security numbers and accounts for all of Iran's population. Bell Helicopters had a large team in Isphahan to train Irani pilots and mechanics, and to make plans for a factory for the assembly of new helicopters.

GE technical representatives were stationed in Iran to train Irani air force mechanics in support of the engines in Irani F-4 and F-5 fighters. The U.S. DOD, it turned out, was not well adapted for quick reaction with spare parts on behalf of external customers. The U.S. Air Force's logistics system was geared to supporting its own tactical squadrons, and allied air forces such as Iran's did not always get priority attention. They could submit their requisitions, but what the system delivered months later might not be precisely what they had ordered. GE proposed that Iran buy its spare parts directly from GE. The quality would of course be the same, and GE would do nothing without an export license from the U.S. government. But it could assure the Iranian air force that it would get the part numbers it had ordered, in the quantity specified, at the promised delivery time, rather than having to accept whatever might trickle out of DOD's long logistics chain.

It was an attractive argument, even if GE may have oversold its capability. But there was a formidable obstacle: the Irani Air Force had very little logistics management capability of its own and, for that, depended largely on the U.S. Air Force. The commanders were reluctant to give up that strong shoulder, or perhaps to give up having someone else to blame if anything were to go wrong. GE needed an intermediary in Iran who could persuade the Irani air force to buy spare parts directly from GE, to make clear that this would in fact improve its logistical situation, that GE was a dependable supplier, and that the DOD would always remain available as a supplier of last resort. In other words, GE needed to find someone trusted by the Irani air force to make the case that the risk was small and manageable. GE itself would be regarded as a biased witness, just as a carpet salesman of Persian carpets may not be entirely trusted by a foreign tourist.

Such a matter required an intimate knowledge of the Iranian air force's logistics system and organization. GE hired an Irani firm, Management & Technical Services Company (MTC), which acted as sales representative for a number of aircraft companies. MTC was headed by Abolfat Mahvi, a man with a strong physical resemblance to the Shah of Iran, a notion he did not play up but did not discourage with his foreign clients. With MTC's and Mahvi's help, the Imperial Iranian Air Force finally undertook to buy engine spare parts directly from GE. They got good service, and GE got good prices. MTC earned a healthy commission from the sales. Suddenly, in the wake of the revelations of Senator Church's committee, the Shah decided to forbid the use of any agents in the procurement of military equipment. All sales had to be made by a government or directly by the principal manufacturer, not by distributors; sales invoices had to be accompanied by a declaration notarized by the U.S. Embassy in Tehran that no agent had been used. GE submitted a declaration that it was using MTC as a sales representative. Other companies made similar declarations. The immediate result was that the Iranian War Ministry issued a "black list" of all such agents, MTC among them, ruling out any further transactions with them, upon which GE at once terminated its agreement with MTC. Mahvi threatened: GE owed him some $1.4 million in commissions on sales already made and he wanted his money. GE acknowledged this fact, and explained to him that as long as the "black list" existed, GE could not legally pay him what MTC had legally earned. He needed to use his ingenuity to get MTC's name off the list. Mahvi did not accept GE's position. However, in a few months the Iranian War Ministry took MTC's name off the list, along with the names of other companies, and GE was able to pay Mahvi the commissions he had earned. But the sales representation agreement was not renewed.

Chapter 12

The Gap in Everyone's Commercial Product Line

THE biggest military aircraft markets in the 1960s and 1970s were those for fighters and attack aircraft and for utility and antitank helicopters; new turbofan and turboshaft engines were developed for them on both sides of the Atlantic Ocean. As national programs grew more and more costly, aircraft manufacturers were encouraged by their governments to look for export sales to help pay for the development; lively sales competitions took place in regions of the world where political tensions justified arms procurement and where the funds could be found either from local resources or the United States, the USSR, France, and Britain playing power politics—in the Middle East, for example, and East Asia. These military sales competitions had their counterpart in the commercial market. Jet travel grew rapidly and began to displace other modes of passenger transport, particularly on transoceanic routes but also on shorter routes in North America. By the early 1970s, the first generation of jet liners, the Boeing 707 and the Douglas DC-8, were beginning to give way to wide-body jets such as the 747 and the DC-10. The new big twin-engine airliners, the Airbus Industrie A300 and the Boeing 757 and 767, were poised to take over the most important intercity routes from earlier jet designs such as the Boeing 727. That left another and very important segment of the airliner market unserved by the most modern equipment.

There was already a huge market in North America and Europe for smaller airliners. Deregulation of the airline system in the United

States led to a route structure of a few "hub" cities for each major airline, linked to smaller cities around them through "spokes." The airlines collect passengers at each hub and move them from one hub to another with wide-bodied airliners, while smaller airliners feed traffic along the spokes into and from the hubs (express parcel delivery services are following the same pattern, invented by Frederick Smith, the founder of Federal Express).

Through the 1970s, the 100-seat Douglas DC-9 and 150-seat Boeing 727 were the dominant airliner models for traffic along the spokes (the British Aerospace BAC 1-11 and the Fokker F50 also found a smaller market, mostly in Europe but with a number in Asia and with secondary airlines in the United States). They and later the Boeing 737s all used the same engine, the Pratt & Whitney JT8D, a low bypass turbofan developed around the core engine of the military J52 turbojet. The JT8D was a dual rotor front fan in the classic pattern of Pratt & Whitney designs and was a phenomenal success. At one point, more than 12,000 were in use on several different aircraft types, so that its fortunes did not depend on one type alone, and served efficiently and reliably around the world. Engine operating data collected by the Air Transport Association allowed a rough calculation of the JT8D's economics: the average airliner flies at least 2,500 hours per year; the JT8D consumed spare parts at a rate of approximately $50 per flight-hour, and the profit margin on such spare parts could well have been 50%. If these numbers are of the right order of magnitude, the JT8D program must have generated gross profits to Pratt & Whitney of not less than $750 million per year.

When three engine models such as the JT9D, CF6-50, and RB.211 compete for the same market, there is savage price competition. Small differences in performance or reliability are equalized through price concessions. The manufacturer with the smallest market share is unlikely to recapture its development costs. The ideal situation from the point of view of the engine manufacturer is to be the only supplier of an engine of a certain thrust class, and the JT8D was in this fortunate position. The market share of the Rolls–Royce Spey in the BAC 1-11 was so small that it made no significant difference to the JT8D's dominance. Even so, Pratt & Whitney could not afford to gouge its JT8D customers on price because they were also customers for other engines in big airliners, and dissatisfied customers are bound to take advantage of alternative suppliers as soon as they can find them. But such a quasi-monopoly position eliminates the price pressure on profit margins for installed engines as well as spare parts. The JT8D was a wonderful money machine for Pratt & Whitney, and that did not go unnoticed by the competition.

In 1968, GE's advanced product planning operation began studying a new high-bypass turbofan engine that would have substantially better performance than the Pratt & Whitney JT8D with its low bypass ratio and comparatively low turbine inlet temperature. GE's plan was to create an engine suitable for the DC-9 and 737, other new twinjets that GE anticipated would be designed for the short- and medium-haul market, and the U.S. Air Force KC-135 tanker, all aircraft requiring an engine thrust of 22,000–25,000 lb. A new engine would have to be markedly superior not only to the JT8D but also to a JT8D or Rolls–Royce Spey refanned to a higher bypass ratio.

For the new commercial turbofan study engine, the GE13, the engineers wanted a core engine more advanced than that of the CF6-50 and had such core engines potentially available in two demonstrator engines for the U.S. Air Force advanced turbine engine gas generator (ATEGG) program, the GE1/9 and the GE1/10. With very advanced compact compressors of nine or ten stages, and with single-stage turbines, their core engines were short enough that they could be supported with two bearings rather than the three of the CF6. One less main engine bearing means one less frame to carry the loads and one less oil sump with oil supply and scavenge pumps—significant reductions of complexity, weight, and parts cost. At the time of these studies, GE was also designing another prototype engine, the XF100, for the U.S. Air Force F-15 fighter competition, with a high stage-loading core compressor and a single-stage turbine and a differential bearing in the rear so that the entire engine would have only two structural frames. For the GE13, the advanced compressors promised excellent performance, and the combination of only two structural frames and two main bearings promised a simple and light mechanical design.

But GE lost the competition for the F-15 engine to Pratt & Whitney, in part because it had no operating experience with dual rotor turbofans with afterburning, while Pratt & Whitney had gained such experience on the TF30. (Much of the experience had been bad. The TF30 was designed by Pratt & Whitney for the U.S. Air Force F-111 and the USN F-14. In the F-111, the engine suffered from severe compressor stalls caused in part by airflow distortion from the aircraft's inlet and, in part, by the fan's sensitivity to such distortion. In the F-14, the engine's thrust simply was not adequate for the U.S. Navy's flight operations from aircraft carriers.) GE switched its XF100 designers to work on the engine for the Advanced Manned Strategic Aircraft, which became the B-1. Here, too, GE decided to go with an advanced core, even more advanced than the GE1/9 ATEGG demonstrator engine. The new engine had a bypass ratio of two, a nine-stage compressor, and a highly loaded single-stage turbine. In 1970, the U.S. Air Force accepted GE's proposal, and the new F101 became GE's first supersonic dual-rotor front fan. The F101 core also fitted the GE13 subsonic turbofan perfectly. Preliminary design studies continued on a 22,000-lb thrust version of the GE13, with exploration of several fan mechanical designs and various nacelle configurations.

While these military competitions played out in the United States, the French government had been supporting Airbus Industrie with financing for airliner development. In parallel, it had set a strategic objective of enlarging SNECMA's role in the civil engine market, to complement its technical capability for developing military engines for French combat aircraft and, if successful, to support the military programs with commercial profits. In 1970, wide-body airliners with high-bypass turbofan engines were clearly beginning to displace an earlier generation of airliners such as the 707 and DC-8. The new president of SNECMA, General René Ravaud, was convinced that a modern high-bypass turbofan engine could also take the place of the JT8D and the Spey in a new generation of smaller airliners. None of the big three engine manufacturers had an engine of this size in development.

SNECMA began design studies for such an engine, the M56, its target application a hypothetical 150-seat airliner, the Joint European Transport 1 (JET 1). The French government supported the concept but recognized that SNECMA lacked the technical resources and reputation with the airlines to launch the M56 on its own. Ravaud was encouraged to seek cooperation from any of the big three manufacturers, Pratt & Whitney, Rolls–Royce, or GE. This came at a time when Pratt & Whitney's parent company still owned 11% of SNECMA's shares, received as payment for licenses on Pratt & Whitney technology.

In contrast with its aggressive marketing efforts in Germany, GE had not made a strong attempt to break into the French military engine market. The French government insisted on its own fighter and helicopter engine developments independent of the United States and, to that end, funded SNECMA and Turbomeca. Instead, GE had entered into a commercial engine collaboration with SNECMA on the CF6-50, whose importance turned out to be much greater to GE than any benefit that could be derived from trying to sell a few military engines to the French government. On the last day of the Paris Air Show in June 1971, GE's Gerhard Neumann visited Ravaud at the SNECMA chalet to introduce himself and pay his respects. Ravaud, learning that Neumann was returning to the United States the next day, asked him to visit him in his headquarters on the way to the airport. There Ravaud raised the subject of the M56.

Pratt & Whitney appeared to be engrossed with the success of its JT8D product line. In fact, Pratt & Whitney was offering a version of the JT8D with a new fan for the 150-seater application, to which the M56 would be a competitor. Rolls–Royce had gone into receivership after the RB.211 fiasco of 1970 and was preoccupied with the problems of supplying engines for the Lockheed L-1011, and in any case Ravaud was not fond of Rolls–Royce, an antipathy strengthened by SNECMA's experience in joint programs such as the Olympus for the Concorde and the M45. SNECMA's experience working with GE on the CF6-50 had been good and promised to be profitable. What would it take, he asked, for GE and SNECMA to work together on the M56?

Neumann replied that the first step had to be for the two of them to come to a complete agreement on such a program. Obviously GE's headquarters would have to approve, but first the two of them had to be of one mind. Neumann said that it would take him about one month to assemble a technical team, that he and Ravaud should talk by phone every week, and that they should meet next in the United States to work out a detailed plan. GE had the GE13 studies, designed around a core engine developed for the F101 turbofan for the B-1 bomber, its core the right size for the M56. Using it would greatly reduce the cost and technical risk of the development program. They shook hands on the deal—or tried to: Ravaud had lost his right arm during the war. Both of them laughed, and shook again with their left hands.

They met a month later at the GE factory in Lynn where Neumann had his office. The SNECMA engineers brought their M56 data, and GE showed them a cross-section drawing of the GE13 and calculated performance data. Over the next few days, Neumann and Ravaud roughed out the elements of a collaborative program, in which the two companies would share equally. They made a crude cost estimate: the fan and its turbine and the gearboxes cost about the same as the core engine; that could be the basis for equal sharing of the work of development and production. GE could work on the core, based on the F101, and SNECMA would be responsible for the fan and fan turbine and the accessory drive gearboxes. They agreed that neither side would ask the other for a cost accounting; the exchange rate between francs and dollars could go up or down, but each side would pay its own bills. Both companies would assemble complete engines, GE making twice as many core engines and SNECMA twice as many fans and low-pressure turbines as they would have done had each decided to build all engine parts itself. SNECMA would be the project leader. Ravaud asked GE to take responsibility for systems engineering and proposed that SNECMA take responsibility for marketing in the United States. Later, GE became responsible for marketing to some airlines where it had particularly strong relationships. (Integrating the marketing work between SNECMA and GE turned out to be more difficult than integrating engineering and manufacturing, where the two sides got along smoothly from the beginning.) The engine would be called the CFM56.

The plan was simple and easy to explain to the boards of the two companies. Neumann and Ravaud got approval to go ahead and formed a joint company, CFM International (CFMI), to act as program manager. It would have only a skeleton staff: all services such as engineering, manufacturing, marketing, and product support would be performed for CFMI by its two owners, with the work balanced to match their equal ownership.

GE at once proposed the CFM56 to the U.S. Air Force for a new military freighter, the Advanced Medium-Range Short Takeoff and Landing (STOL) Transport (AMST). Because SNECMA would pay for its half of the work, GE could propose a development bill to the U.S. government much smaller than if GE had planned to do all the work itself, and the use of the F101 core would reduce development and production costs even further. Of course, to develop the engine for the AMST and then to obtain a civil type certificate, GE and SNECMA would have to exchange some data and hardware for testing.

There were skeptics inside GE and SNECMA who might have preferred another partner, but a bigger obstacle arose immediately from the U.S. government. The U.S. Air Force was opposed to the notion of a foreign partner for engine development, even if there were a big benefit from lower development costs. The F101 core proposed for the CFM56 had been developed for the B-1 bomber, and the DOD was concerned that the engine technology of the most advanced American strategic bomber could leak through France into the Soviet Union.

An exchange of technical data between GE and SNECMA required approval under the U.S. government's export control regulations. The DOD wanted to limit GE's technology level in the program to that of the CF6-50, one step behind the core of the F101 bomber, to keep U.S. engine technology one step ahead of that of any other country, even friendly ones such as Britain and France. That would have limited the performance advance of the CFM56 engine over the JT8D. The problem was not resolved between GE and the DOD, and indeed not until Presidents Nixon and Pompidou met in Iceland. There were many issues on their agenda: the United States was pulling out of Vietnam, there was the question of tariffs on aviation imports to France, and many other political matters. SNECMA and the French Ministry of Transport had put the issue of CFM56 data export approval in President Pompidou's briefing notes, and GE and the DOD furnished briefing notes to the White House for

President Nixon. At the very end of their meeting, the two presidents gave approval to the CFM56 program, provided certain safeguards were put in place. GE would have to be the project leader, technical data from GE to SNECMA would be limited to the interface between the two halves of the machine, and no data could be transmitted on the inner workings of the core engine. Core engines sent to France for engine testing would have to remain under GE's physical control, guarded in a separate protected storage area. (GE heard later that both presidents were so tired by then that they may have paid less than full attention.)

GE and SNECMA began development work at once. The engineers of each company had a good opinion of their own strengths, each quite ready to tell the other team how to do its work. But Neumann and Ravaud insisted on results and would accept no excuses or complaints about lack of cooperation. "Bring us solutions, not problems," the two insisted. Their example quickly pervaded the entire American and French organizations, whose cooperation became remarkably smooth and effective. GE and SNECMA put the first engine to test in 1974, while U.S. Air Force technical experts monitored the progress of engine development in both countries. As time went on, the French became more acceptable to the DOD, and data restrictions were gradually relaxed.

CHAPTER 13

■

But Will It Sell?

GE had the CFM56 marketing strategy all worked out. Development would proceed with engines built around cores derived from the F101 engine program for the B-1 bomber; the U.S. Air Force would buy the initial batch for the AMST. Early engines are always expensive to manufacture, but that would matter less because the U.S. Air Force would audit costs and pay whatever they were as long as they were reasonable and defensible. As production went along, costs would come down, and subsequent engines could be offered to the U.S. Air Force—and of course to airlines—at a lower price, a price per pound of thrust comparable to that for bigger engines such as the CF6-50. The strategy was reasonable, but it turned out to be wrong on every count.

First, President Carter canceled the B-1 program as part of the calculus of the Strategic Arms Limitation Treaty, after GE had already delivered 44 engines to Rockwell. The engines performed flawlessly in the B-1, but the military program on which GE and SNECMA had counted for core engines for the CFM56 had disappeared. To make matters worse, the AMST program never came into being in the form visualized by the U.S. Air Force, eliminating the buyer for the initial high-cost engines. Then came the OPEC oil embargo of 1973, lending emphasis to the development of more fuel-efficient engines but also causing a worldwide recession and slowing down orders for new airliners.

There were many occasions when GE would have wanted to slow down or even terminate the CFM56 program, which was swallowing money at a relentless clip. But a program approved personally by the presidents of two countries is not easy to kill (in early 1979, GE did kill the CF6-32 program, which had SNECMA, Volvo, and Alfa Romeo as partners, after the Pratt & Whitney PW2037 and the Rolls–Royce RB.211-535 engines won the bulk of early orders for the new Boeing 757 at cut-throat prices; GE and its partners reluctantly conceded that it made no sense to go on with a losing proposition). Ravaud and Neumann shared the tenacious conviction that the market for the CFM56 was bound to come. GE's vice chairman, Jack Parker, also believed that there would be a market for thousands of CFM56 engines. There was a clear need to compensate for the increased price of jet fuel and growing public concern about noise and exhaust emissions, and there was no other modern competitor on the horizon. Parker gave the CFM56 program unflagging support at GE's corporate headquarters, protecting Neumann against anyone who might have wanted to divert the investment to other products. At the eleventh hour, the first CFM56 order came in—from United Airlines to reengine DC-8 aircraft.

The DC-8 airframe had been designed for a very long service life and had a very efficient lift-to-drag ratio. Airlines liked its efficiency and passengers liked its comfort. Given the large increases in the price of jet fuel, it made economic sense to replace its JT3D engines with the much more efficient CFM56, at the same time benefiting from the reduced takeoff noise and the big increase in aircraft range. United, then Delta, and then Flying Tigers Airlines all made the conversion—a total of 110 aircraft. It was a good start for the CFM56.

Years before, when Douglas launched the DC-10, it made a fatal decision, scrapping design data and tooling for the DC-8 to avoid stealing sales from the DC-10. Now CFMI faced the problem of getting FAA certification for the CFM56 on the DC-8 without the help of Douglas. With CFMI's support, Jackson McGowen, long retired from Douglas, assembled a small team of other retired Douglas engineers to manage the certification of the reengined DC-8. GE and SNECMA advanced McGowen the money for the certification work and were repaid out of the selling price of each aircraft delivered. When the McGowen team was done, the new DC-8 Series 70, as it was called, went into service with United, Delta, and Flying Tigers Airlines and later with United Parcel Service. The Saudi businessman, Adnan Khashoggi, also bought one and had it equipped as a luxurious long-range executive aircraft. The DC-8 program established a price for the CFM56 in the market: the airlines would not pay more per pound of thrust for a 22,000-lb CFM56 than they would pay for a 50,000-lb CF6-50. But the engine manufacturing cost was still high. Only an increase in production volume could solve this problem. The Boeing 707 was an obvious target.

Ravaud visited Seattle to meet "T" Wilson, the Boeing chairman. Ravaud asked what it would take to put the CFM56 on the 707 airframe. Boeing had little interest in a program that might help SNECMA and GE sell engines but would not generate the sale of new 707 aircraft; the best prospect, Wilson said, was the Boeing KC-135 tankers used by the French Air Force to extend the reach of its *force de frappe*, the strategic bomber fleet of Mirage IVs. When he got back to Paris, Ravaud persuaded the French government that this was well worth funding: it would extend the capability of its strategic bombing force and open a potential market for CFM56 engines, half of it served by SNECMA and its French suppliers. Boeing would hardly refuse such an important customer. The French government agreed and funded the French air force to have Boeing convert its tankers to the CFM56, a sale that paid the nonrecurring cost of designing the CFM56 installation on the 707/KC-135 airframe and the flight tests—comparable to what CFMI had been forced to do with McGowan's team on the DC-8. On the strength of the example of the French air force tankers, GE and SNECMA then persuaded the Royal Saudi Air Force to put the CFM56 into its E-3 AWACS aircraft.

Impressed by the fuel savings demonstrated by the DC-8 and the French KC-135, the U.S. Air Force then decided to reengine its own Strategic Air Command KC-135 tankers, a vital part of the U.S. Air Force's capability for what was called force projection during the Cold War. They were still equipped with their original J57 engines, already expensive to maintain, noisy on takeoff, and with high fuel consumption. Installing a modern high-bypass turbofan would much reduce the noise problem and allow two tankers to deliver as much fuel as three equipped with the J57. The CFM56 was ideal for this application, but there was strong competition from two Pratt & Whitney engines, the commercial JT8D and the military TF33, the engine used in the B-52H bomber and the C-141 freighter. The JT8D was in production at a very competitive price, lower than that of the CFM56, and the U.S. Air Force had surplus TF33 engines in its own inventory. Both engines were all-American, not half French like the CFM56.

Winning the competition was obviously very important for GE. It would enlarge the CFM56 production base so that costs could be driven down enough to put the program in the black, and having the CFM56 in U.S. Air Force inventory would enhance the opportunity for other GE engines using the same core, such as the F101 and its fighter derivative. The U.S. Air Force's basis for selecting the winning tanker engine was the lowest life cycle cost for the entire program: engine price, aircraft conversion, maintenance costs, and fuel. GE planned to offer an attractive if risky low commercial flat price for the CFM56 and believed that its life cycle cost would be a winner. Having lost previous competitions when the U.S. Air Force evaluators used their own methodology for calculating life cycle cost, GE asked for a final opportunity to explain what made its CFM56 proposal so attractive. The U.S. Air Force source selection team declined with thanks: it had all the data it needed from Pratt & Whitney and GE to make its own life cycle cost calculations.

Jim Krebs, in charge of GE's military engine operation, was a strong believer in doing everything one can think of to win—far better than losing and later wondering whether one might have changed the outcome by doing something extra. Krebs called the Judge Advocate General (JAG), DOD's top lawyer, and told him that both GE and Pratt & Whitney deserved a final chance to make their case, whether U.S. Air Force program managers wanted to have more data or not. The JAG must have talked to the Air Force because Krebs was invited to meet with Lt. General Larry Skantze, the commander of U.S. Air Force Systems Command, who was in charge of source selection. They talked for an hour, and Krebs showed the general a photo he had taken on vacation in England near an air force base. The townspeople had put up signs all over town warning "DANGER! The KC-135s are COMING!" Skantze laughed. As Krebs left, four Pratt & Whitney people were waiting for their turn with the general. The U.S. Air Force source selection board picked the CFM56-2 engine for most of its tankers and existing TF33 engines already in its inventory for a few of them. It was an important victory for GE and for CFMI in terms of sales of installed engines. The U.S. Air Force tanker program enabled the CFM56 to reach the break-even point, with total orders for more than 2000 engines. Production rates rose and GE began a rigorous cost-reduction program, with the goal of bringing the cost of the engine down by a factor of three.

The U.S. Air Force had decided some 15 years earlier not to fund development of any more transport engines, on the theory that commercial engines designed to satisfy airlines should also satisfy military transport requirements. The CFM56 was the first engine bought under these "commercial" rules. There is a big difference, however, between the military and commercial purchasing processes. The U.S. Air Force was accustomed to cost reimbursement contracts. Early engines cost a lot, and the U.S. Air Force is used to paying these costs; as costs are reduced with continued production, the U.S. Air Force negotiates lower prices. In contrast, prices for commercial engines are set by competition in the marketplace, whether they cost the manufacturer a lot or a little. The U.S. Air Force liked the low commercial prices set in the DC-8 program for its CFM56 engines, flat prices that eliminated the need to pay more for early production models. The U.S. Air Force was also used to paying separately for the cost of technical manuals and service engineering support, and was delighted to get all of that at no separate charge, just like airlines. But with commercial engines, revenue from spare parts sales pays for technical publications, service engineering, the fixing of problems, and the development of the engine, and the U.S. Air Force was stunned by the commercial price for spare parts.

In an airline, the same purchasing manager buys the engines and subsequent spare parts. The military procurement process separated these functions: the program manager in U.S. Air Force Systems Command negotiated the order for engines and became a hero if the price was low. Another buyer in an entirely separate organization, Air Force Logistics Command, had the responsibility for buying spare parts and technical support. He got no credit for the low engine price, did not understand the economic logic on which commercial spare parts prices were based, and had certainly never budgeted for them. GE helped the Air Force with a compromise. When CFM56 parts were common to those of the military F101 engine, GE charged the U.S. Air Force its "military" price, with audited costs and a negotiated profit. For all the other parts, U.S. Air Force agreed to pay the commercial price. Because tankers fly only about 650 hours/year, compared with the 2500–3000 hours/year of a big airliner, annual spare parts consumption will be low. The U.S. Air Force expects the average CFM56 engine to stay installed for 16,000 operating hours, almost 25 years! If the engineers were right about low spare parts consumption, the pricing formula was not going to make a fundamental difference to program costs for the U.S. Air Force, nor to profits for CFMI. The CFM56 received the type designation F103 from the U.S. Air Force and proved phenomenally reliable and fuel-efficient in service. Noise complaints around tanker bases also became less frequent. The program was a great success.

In 1981, Boeing launched a new version of its 737, with the CFM56 engine (see Fig. 6). There was consternation in France (and in Japan, which had hoped to get the Japanese/Rolls–Royce RJ500 engine into the 737). Airbus Industrie wanted to launch its own 150-seater and was outraged that SNECMA and GE would support another Boeing product, although SNECMA's Communist labor union told the French government that it was better for France to have half of the engine on a Boeing winner than all of the engine on a European airliner with uncertain prospects. Despite that, Airbus Industrie counted for more in French government industrial policy than did SNECMA, and the French Transport Minister demanded to know why GE and SNECMA were favoring Boeing over Airbus Industrie. GE addressed the issue head-on. The French Embassy in Washington, D.C., arranged a meeting for Ed Hood, now a vice chairman of GE, and Brian Rowe, in charge of GE Aircraft Engines, with the French Minister of Transport and the prime minister. Hood pointed out that GE had supported Airbus Industrie, Douglas, and Boeing impartially and that GE's work with SNECMA and Airbus Industrie had played a major role in the market success of the A300. Would GE support Airbus Industrie on a new 150-seater, Hood was asked?

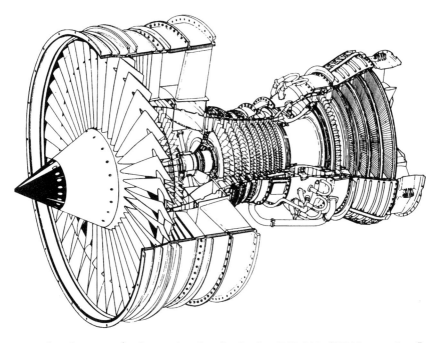

Fig. 6 **The CFM56—the engine for the Boeing 737-300.** (CFM International)

Of course. Would GE develop a new engine for such an aircraft? Hood and Rowe replied that the market may not be big enough to justify developing a completely new engine; the CFM56 would serve the purpose well and offered minimum risk and maximum economic benefit to SNECMA and GE. Would GE assist SNECMA if the French government decided to launch a new engine development program itself? Yes, GE would do that if it were funded for its share of the work, as indeed SNECMA would be. But, in that case, it would be important, GE said, to launch the new aircraft right away, so that it would not be left behind in market share.

Three years later, in 1984, Airbus Industrie launched its own new 150-seater, the A320, for which SNECMA and GE agreed to develop a new version of the CFM56 with a smaller fan. In effect, both the new 737-300 and the A320 became the 150-seater JET1 originally envisaged by Airbus Industrie in 1970 at the time René Ravaud started the M56 studies at SNECMA. Since that time, Airbus Industrie has launched other versions of the A320, and other airliners using the CFM56: the larger A321, the smaller A319, and the A340, a four-engine long-range aircraft. Boeing is also delivering a new series of 737 models with a new wing and an advanced model of the CFM56. More than 10,000 CFM56 engines have been ordered so far, and GE expects the program to be as big as the Pratt & Whitney JT8D was in its day.

The CFM56 has had an important technical and economic impact on GE's military engine programs and vice-versa. B-1 production was reinstated by President Reagan so that the CFM56 core found the military application for which it had originally been designed. Later, a highly modified version of the F101 was used for the B-2 stealth bomber. The same core with a smaller fan and different afterburner became the F110, the fighter engine that replaced the Pratt & Whitney F100 in U.S. Air Force F-16s and the Pratt & Whitney TF30 in U.S. Navy F-14s. The air forces of Turkey, Israel, Greece, Egypt, and finally Japan, followed the lead of the United States and also selected versions of the GE F110 for their fighters. These customers' good experience during years of operation with GE's earlier fighter engines, the J79 and J85, and GE's product support, also contributed, no doubt, to the choice of the GE F110.

Pratt & Whitney fought hard against this unexpectedly effective competition from GE. It had studied a high-bypass turbofan in the 25,000-lb thrust class, the JT10D, in collaboration with Rolls–Royce, a program that came to nothing, in part, because of the fierce competition between other commercial engines in their product lines and, in part, because Pratt & Whitney had put its bets on the JT8D program's permanent market monopoly. Pratt & Whitney also believed that GE could not pull off a successful collaboration with SNECMA, with which Pratt & Whitney had long-standing agreements to exchange technology. The CFM56 came as a stunning surprise.

What astonished Pratt & Whitney most was GE's competition for the F-16 fighter engine. GE's F110 engine was immediately competitive in price with the Pratt & Whitney F100, even though Pratt & Whitney had delivered more than 2000 F100 engines to the government. The secret lay in the simpler mechanical design of the GE engine, and in the cost reductions inherent in the large production program for multiple applications: GE used the same core in engines for airliners, for two bombers, and for supersonic fighters. Technical improvements developed for airline engines could be fed back into the military programs and vice-versa. Evidence is hard to come by, but there is reason to believe that, in the 1980s, GE's production costs were the lowest in the industry, for both its commercial and military engines.

Chapter 14

The Japan Factor

There is always competition. Against the new 737 and A320, Douglas managed to keep the DC-9 design alive as the MD-80 series with refanned versions of the venerable Pratt & Whitney JT8D. But a more interesting challenge to the CFM56 came from Japan. The Japanese—an amorphous concept that embraces the government, its economic planners, and Japanese industry—decided to become global competitors in the aircraft engine business.

GE had a significant business presence in Japan with steam turbines, generators, electrical machinery, and light bulbs since the 1920s, including a major share holding in the Tokyo Shibaura Electric Company, now known as Toshiba. During the Korean War, the Japanese engineering company Ishikawajima–Harima Heavy Industries, Ltd. (IHI), began to make J47 parts. In 1960, IHI began to manufacture J79 engines under GE license for the Japan Air Self Defense Force's F-104J, and that collaboration was subsequently extended to engines for F-4J Phantoms, T58 helicopter engines, and T64 turboprops for Japanese P-2J maritime patrol aircraft built under a Lockheed license. Despite its long experience in Japan on such programs, GE found it difficult to penetrate the Japanese commercial engine market. There had been no significant commercial aviation in Japan before World War II. After World War II, Japan Airlines (JAL) was set up under the MacArthur occupation as a government-owned airline with Northwest Airlines as its mentor. Northwest used Pratt & Whitney engines and, therefore, so did JAL. Pratt & Whitney was represented in Japan by Mitsubishi Trading Company, many of whose managers had close personal relationships with JAL managers dating back to their days as classmates in Tokyo University. The relationship was easy, and JAL enjoyed a good product and good support from Pratt & Whitney.

In the mid-1960s, JAL ordered the Convair 880 for its north–south routes to Taipei and Hong Kong. Its experience with the CJ805 engine was not entirely happy. Japan Airlines judged the engine less sturdy than its Pratt & Whitney engines and was disappointed in GE. GE believes to this day that it gave JAL good support, but this was GE's first exposure to Japanese quality expectations, and no doubt there were some mutual misunderstandings. In any case, JAL's senior managers were not satisfied with the CJ805 nor with GE, even though the Convair aircraft continued in service for a number of years. Whether GE realized it at the time, in the classic Japanese manner JAL's market was closed to GE *as long as its senior managers were alive* because, even after they had retired, some of them would continue to serve the airline as advisors on equipment selection. Until a completely new generation of management was in charge, GE would in effect be shut out of JAL.

When GE launched the CF6 engine, it decided to reenter the Japanese market through two new carriers, All-Nippon Airways (ANA) and Toa Domestic Airlines (TDA). Both were privately owned and more flexible in their management attitudes than the government-owned JAL. Both also wanted to take market share away from JAL and were open to any opportunity that might help them compete. GE offered CF6 engines to ANA and TDA on attractive terms, assuring them that GE's product was as reliable as Pratt & Whitney's and that GE had learned how to support Japanese customers. Besides, GE noted, Pratt & Whitney was no doubt committed much more to JAL than to upstart competitors. The strategy was successful. ANA bought a special version of the 747 for short-range routes inside Japan, with CF6 engines de-rated to 45,000 lb of thrust. Toa Domestic Airlines bought Airbus Industrie A300B aircraft with CF6-50 engines. GE put a team of field service engineers into Japan to support the engines and smothered both airlines with attention. Both airlines were successful in enlarging the market and taking a share of what had been a JAL monopoly when the market was smaller. CF6 engine operation was exemplary. Japan Airlines could not help but notice and when, in a few years and with a new generation of managers in charge, JAL held an engine competition for its new big 747-400s, the record was there: the CF6 service experience at ANA and TDA and GE's exemplary product support were sufficient for JAL to select the GE CF6-80C2 engine. Competition between engine makers, it must be noted, is good for the airlines. For its Boeing 777, JAL went back to Pratt & Whitney for the PW4000 engine.

Japan is a small country, with no long-range internal routes. But as an island nation with a focus on world trade and with high national income per capita, Japan is an important market for aviation products. Japan Airlines alone has about 100 Boeing 747s; ANA ordered 40 Boeing 767s at one time. Altogether the three Japanese airlines represent 3–5% of the world market, with Japan importing almost all of its commercial aircraft from the United States or Europe, American civil aviation sales to Japan help to offset the large imbalance of trade between the two. Unlike some other countries, Japan has not insisted that foreign manufacturers of commercial aircraft make purchases in Japan to "offset" the effect of the import, despite the Japanese government's emphasis on creating a Japanese airframe and engine industry capable of competing on the world market. Unlike the situation in other industry sectors in which Japan has developed a strong competitive capability and market share since the end of World War II, the world's airframe and aircraft engine markets continue to be dominated by American firms, with strong European competition but little from Japan. The situation differs so radically from that in other industry sectors (in steel, automobiles, consumer electronics, semiconductors, construction machinery, and photographic film, to cite only a few prominent examples) that the reasons are worth a detailed examination.

Aerospace development requires very large capital investments. American industry and the U.S. government have been making such large continuing investments in research and development for more than 80 years, giving the United States a big lead over any new entrant. The Cold War gave the United States the incentive to keep advancing its military aviation technology, and much of this could be spun off into commercial aviation. In contrast, Japan for many years stood apart from the development of military technology, concentrating on commercial products and being content to obtain military technology as required from the United States.

Although the Japanese civil aviation market is substantial in absolute terms, it is not large enough by itself to justify the initial capital investment for development of an airliner. American domestic demand in contrast is as much as 10 times larger. The American domestic market is not closed to imports but nevertheless prefers domestic products for reasons of proximity and easy access to product support, leavened perhaps by patriotic sentimentality. In any case, the opportunity does not exist in aviation to develop a product, refine it, and bring its costs down in an internal Japanese market that rejects competition from imports, and then exporting on a price-competitive basis.

The American industry paid strict attention to the economics, quality, and reliability of its products—in most ways they set the world standard. Its costs are among the lowest in the world because of high productivity and large production runs. The industry is very sensitive to the demands of the market and attentive to the needs of its customers. Most of all, the industry and the U.S. government remain hypersensitive to any potential assault on the American comparative advantage from Europe and Japan.

Since the end of World War II, the Japanese aviation industry has not managed to make a success of developing its own commercial aircraft, despite considerable prodding by and financial support from the government. The one major national program, which involved all the airframe companies in association, designed and produced the YS-11 twin-turboprop airliner, with two Rolls–Royce Dart engines. The YS-11 entered the market late, in competition with other established turboprop airliners such as the Lockheed L-188 Electra and the Vickers Viscount, and in the face of overwhelming competition from turbofan-powered airliners such as the DC-9 and 737. The YS-11 was not a bad aircraft—its performance was about as advertised—but it was not significantly better than any of its competitors. It was late and its cost was high because of limited production quantities. The program was halted after production of fewer than 200 airliners.

Later in this chapter, we discuss the steps taken by the Japanese engine industry, supported by its government, to establish a competitive capability in the field of commercial engines. As we shall see, this led to an initial failure, after which the Japanese elected to develop strong niche capabilities and play the role of junior partners to American and European project leaders. Kawasaki, Mitsubishi, and Fuji Heavy Industries have participated in production of the Boeing 767 and in the design and production of the Boeing 777. Kawasaki, Mitsubishi, and Ishikawajima–Harima also participate in the development and production of the latest turbofan engines for aircraft such as the Airbus Industrie A330 and the Boeing 777: the Rolls–Royce Trent, the Pratt & Whitney PW4000, and the GE90.

Why, then, did the American prime manufacturers offer production shares and, more recently, development shares, to the Japanese airframe and engine companies? Are they not risking that the Japanese companies will learn enough from such cooperation to catch up and surpass them in one of the few industrial sectors where Americans and Europeans enjoy a comparative advantage? There is a risk of this happening, but the reasons for including Japanese partners in the development and production of airframes and engines are plain: the Japanese have carefully educated themselves to create a strong specialized capability, particularly in advanced manufacturing technology. They do work of high quality; they finance their share of the work; and they accept prices at competitive world levels for what they do, confident that with a share of the worldwide market they will be able to bring their costs down below the price. That makes the Japanese manufacturers attractive partners. If Japanese participation in a specific program enhances its market opportunity in Japan, so much the better.

Sometimes Japanese participation appears to help in the market and sometimes not. When ANA bought 40 wide-body twin-engine airliners, the competition was between the Boeing 767, with substantial Japanese content, and the Airbus Industrie A310, with none. The airframe companies will argue that the performance of the two is about the same, and no doubt the price was substantially the same also. All-Nippon Airlines selected Boeing. On the other hand, when ANA later bought the Airbus Industrie A320, the engine choice was between the GE/SNECMA CFM56, with no Japanese content, and an engine in which all three Japanese engine companies participated with Japanese government funding. All-Nippon Airlines stubbornly picked the CFM56 engine.

The activities of the Japanese engine industry had their source in a decision of the early 1970s, when the Ministry of International Trade & Industry (MITI), the postwar successor to the Munitions Ministry set up in 1943 to organize national production during World War II, set up a series of task forces of bureaucrats, professors, and industrialists for advice on appropriate directions for the Japanese economy. Economic planning and industrial policy are honored concepts in Japan. The advisors and the ministry agreed that cameras, ship building, and steel making had run their course. The latest plant would always have the lowest marginal cost and could, therefore, charge a lower price; newly industrializing countries such as Brazil, Korea, and Taiwan were building shipyards and steel mills with exactly those objectives. In contrast to the British government, which subsidized British shipyards to build ships for Poland at a loss, the Japanese government encouraged its firms to reduce the size of their yards. Ishikawajima–Harima Heavy Industry, Ltd., the world's biggest ship builder, with a production that exceeded that of all German shipyards combined, decided to move some of its ship building to a joint venture in Brazil. Nippon Steel cut apart an entire steel mill and shipped it to China, where it was welded together, giving the Chinese a modern addition to their steel-making capacity much faster than it could have been built from scratch. The MITI study concluded that Japan should concentrate on economic sectors with high added value and technology content, where it would encounter less competitive pressures from emerging economies such as Korea and Malaysia. Aircraft and aircraft engines were two of the sectors selected. Although MITI counted on getting technology and information from outside Japan, MITI also concluded that if Japan wanted to be taken seriously as a participant, it would have to develop its own technology and capability in the field. The example of the YS-11 airliner showed that the task was more difficult than had been the case with cameras or automobiles. The world's airlines were not ready to accept Japan or anyone else as a major supplier without convincing proof that their product would be economical and reliable and their product support as good as that of the Americans and of Airbus Industrie. Nor were the American and European airframe manufacturers complacent about product quality, price, and competition, as the automobile industry may have been. Their size dwarfed that of the Japanese industry. Growth demands enormous capital investment and a continuum of research and development funding.

The Ministry of International Trade & Industry and Japanese firms drew the logical conclusion: Japanese firms would have to bring themselves up to world levels of capability in component and subsystem design and, above all, in manufacturing technology. They met the challenge, with seed money from the government and on occasion with program funding from a government financial agency set up for that purpose. Then they began to play a role in major airframe and engine programs led by the major systems firms in the United States and Europe, with the objective of taking a proportional share of the prime contractors' worldwide programs.

Military Aircraft Production in Japan Sets the Stage

At the end of World War II in 1945, the U.S. occupation exercised a profound liberalizing influence on Japanese politics and the economy. A constitution more or less imposed on Japan transferred power from the emperor to the citizens, and the Diet (the system of aristocratic land tenure) was ended and ownership of land was devolved to the farmers who worked on it; women received the vote; the large industrial trusts, the *zaibatsu*, were substantially broken up. Gen. MacArthur prohibited all activities related to aircraft in occupied Japan. The Japanese aircraft engine industry ceased to exist. Some of the engine companies such as Nakajima simply disappeared, their personnel absorbed by other industries. Several Nakajima engineers landed at Ishikawajima–Harima Heavy Industries, Ltd., Japan's largest shipbuilder and manufacturer of engineering products such as bridges, cranes, tunneling machines, and hydraulic equipment.

But as time and the Cold War went on, MacArthur's preoccupation with communism began to change the direction of these liberal occupation policies. When the Korean War broke out in 1950, MacArthur quietly arranged for logistic support in Japan for the allied forces. Several Japanese firms were mobilized in 1952 to do aircraft engine repair and overhaul for the U.S. Air Force and, as time went on, to manufacture spare parts. In 1954, the Japanese national police was converted into a much larger "Self Defense Force," including a Japan Air Self Defense Force (JASDF). Resurrected Japanese engine companies began to build engines for the JASDF. Three firms constituted the bulk of the aircraft engine business and do so still. The largest is the aerospace group of Ishikawajima–Harima Heavy Industries, Ltd. (IHI), which is the largest because its then chairman saw aircraft engines as an important opportunity and invested to an extent considered uneconomic by his peers. It employs about 3000 people to work on aircraft engines, about 60% of total Japanese capacity. The aerospace group of Kawasaki Heavy Industries, Ltd. (KHI), is the second largest, with about 2500 people working on aircraft engines and about 25% of total Japanese capacity. KHI also manufactures ships and machinery and has an aircraft design and manufacturing department. The smallest of the three is the aircraft engine department of Mitsubishi Heavy Industries, Ltd. (MHI), with about 1500 employees and 15% of total capacity. Another division of MHI is Japan's largest airframe manufacturer. IHI is the only one of the three not involved in airframe design and manufacture.

Reestablishing the industry looked to many like a return to the military production of the defeated imperial government. The Japanese government continued to exercise strict control over the industry, under the supervision of the occupation authorities to which any political blame could be passed. A law passed in 1954 limited participation in the industry to those companies authorized by the Japanese government. There was certainly no room for independent companies to enter the business, although Fuji Heavy Industries (FHI), the lineal descendant of the World War II Nakajima Aircraft Company, at one point tried to set up a small division for engines. Later, FHI again became involved in airframe manufacture.

The engine industry's only customer was the Japan Defense Agency, sometimes acting as an agent for the U.S. government for overhaul and repair but mostly buying equipment for the new JASDF.

As time went on, IHI built military aircraft engines for the Japanese government under license from GE and later from Bristol–Siddeley, Allison, Turbomeca (Turbomeca was the outsider in the French aerospace industry, privately owned; the French government was its largest customer, but Turbomeca acted with great independence at the instinct of its president and biggest shareholder, Josef Szydlowski), and Pratt & Whitney. Such "national" programs involved all three companies. The IHI chairman, convinced of the industrial importance of gas turbines, was prepared to subsidize IHI's aircraft engine division with profits from its big shipbuilding division. Both KHI and MHI saw bigger opportunities for profit in other activities and ceded leadership to IHI. Thus, IHI became the prime contractor for most of the engine programs, but typically KHI received about 25% of the work and MHI received about 15%. On some engines, for example, the Lycoming T53 and T55, KHI was prime; on the Allison T63, it was MHI. All three companies assigned design and manufacturing engineers to these national programs and learned the basics of manufacturing jet engines from their foreign licensors.

Commercial Engine Experiments

One of the engines IHI was building was GE's T64 turboprop, installed in a Japanese version of the Lockheed P-2 maritime patrol aircraft built under license by Kawasaki; it was also installed in a unique military four-engine flying boat, the Shin Meiwa PS-1, which used a fifth engine, a GE T58 built by IHI, to drive a compressor for boundary-layer blowing—a device that allowed the PS-1 to fly very slowly without stalling. IHI planned to use the T64 as a starting point for commercial turbofan engine development. With some data from earlier experiments by GE, IHI built and tested a demonstrator front fan driven by the T64 gas generator. Once IHI demonstrated that it was capable of designing such a fan, MITI funded a

"National Engine Development Program" for a flightworthy turbofan, the FJR710, with a takeoff thrust rating of about 12,000 lb. Under IHI's leadership, the three Japanese companies designed what looked like a scaled-up version of the T64 turbofan demonstrator and tested it successfully on the ground. But Japan lacked a test cell capable of simulating altitude flight conditions. The USSR had then the world's largest altitude test facility. Japan's mutual security treaty with the United States would have made it very difficult for the Japanese to cooperate with the Soviet Union on an advanced technology program with obvious military as well as civil application, and so IHI as the project leader went shopping for help from other countries where such altitude test cells existed: France, the United States, and Britain.

The French government consistently regarded Japan as a dangerous potential economic competitor. If there was any discussion of testing the FJR710 in France, it would almost certainly have come to nothing. IHI did ask GE whether it would make its altitude test cell available. GE was used to working with IHI but could not convince itself that it had anything to gain from helping Japan to develop the FJR710 and instead offered to work with IHI on a new engine built around an existing GE core, for example, that of the new TF34. There would be a clear benefit to GE as well as the Japanese. But that did not fit the Japanese objective of developing its own engine. IHI then went to Britain, and the British said yes without hesitation, offering to test the FJR710 in the wind tunnel of the Royal Aeronautical Establishment at Farnborough.

On behalf of the national engine program, IHI accepted. The three Japanese firms designed and built their engine. The FJR710 met its performance specifications and a few were produced and installed in an experimental Kawasaki transport aircraft. The four engines blew their exhaust over the upper surface of the wing, producing lift at slow flight speeds by the so-called Coanda effect (first described by the Romanian aerodynamicist Coanda) and applied in other experimental transports capable of short takeoff and landing such as the Boeing YC-14 and the Antonov An 34. But the FJR710 engine was never taken through a full type certification, because the Japanese had their sights on a different market.

The Real Product

The MITI aerospace task force was led by Dr. Osamu Nagano, the retired head of IHI's aerospace division, and a distinguished engineer who helped develop Japan's first jet engine for the Imperial Navy at the end of the war. Dr. Nagano's committee concluded that there was no point in competing directly with the world's established engine firms with a big engine. Pratt & Whitney, GE, and Rolls–Royce each had such engines in the market already, and there was no reasonable expectation that Japan could overcome that lead. Further, there was little purpose in developing a direct competitor to the CFM56, at that time under development at thrust ratings of 22,000–25,000 lb.

Like the French government and SNECMA's René Ravaud, Nagano believed that the market would ultimately need a modern replacement for the Pratt & Whitney JT8D, at about 16,000-lb thrust, and for the Rolls–Royce Spey at 14,000 lb. Nagano's committee recommended that MITI sponsor the development of such an engine, in a range of thrusts below that of the CFM56. Japan certainly had the financial resources to do so, and the level of technology acquired from its licensed production of foreign engines and the National Engine Development Program. The Japanese government agreed and created elaborate procedures to support a national development program that would avoid the appearance of direct government funding, for example, loans from a national fund whose resources came from the treasury. The Japan Defense Agency also paid exceptionally high prices to the Japanese engine companies for military production, with the explicit understanding that the firms would invest some of their profits in the National Engine Development Program. Including tooling and facilities, the government paid prices for the aircraft and engines produced under American license twice or three times as high as it would have cost to import them from the U.S. licensors.

As a condition for its support, MITI directed the Japanese engine industry, which had yet to earn a reputation in the world market for civil aviation, to take a foreign partner with that reputation. Nagano responded that neither Pratt & Whitney nor GE would be willing to reveal secrets of design and systems management and, even if they were, the U.S. government would restrict what technology they would be allowed to transfer to Japan. In fact, the U.S. DOD had in 1972 put just such restrictions on the transfer of information by Pratt & Whitney to IHI for the licensed manufacture of the F100 engine in the F-15J fighter as well as on the Allison T56, a turboprop engine of comparatively outmoded technology used in the P-3J maritime patrol aircraft. These were the first such U.S. government restrictions, nothing of the kind having been applied to earlier manufacture of the J79, T58, or T64 engines by IHI under license from GE. DOD intended to keep advanced military technology in American hands, but the Japanese saw it as a mechanism for protecting the U.S. commercial advantage.

The British, in Nagano's opinion, were not in a position to be so fussy. Rolls–Royce had gone into financial receivership in 1971 because of the problems of the RB.211 certification and was struggling to stay alive in the cut-throat competition for big engines with Pratt & Whitney and GE. Rolls–Royce might be amenable to an engine development program in which the Japanese would be treated as a nominal partner, even if the relationship were more that of apprentice and master. The three Japanese engine firms formed a separate company, Japan Aero-Engines, Ltd. (JAEL), headed by Dr. Kaneichiro Imai, the general manager of IHI's aerospace group, who had been Nagano's assistant at Nakajima Engine Company during the war. JAEL entered into an equal partnership with Rolls–Royce to develop a new commercial turbofan of about 15,000 to 20,000-lb thrust, the RJ500. Rolls–Royce at that time had more than 40,000 employees while the three Japanese companies had a combined work force of fewer than 3000. The Japanese simply did not have the manpower to do half of the development and production work. Japan put up its 50% of the project funding but much of the work had to be subcontracted to Rolls–Royce, which alone had the capacity to do it. Nevertheless, the Japanese got their money's worth. At any one time, some 90 Japanese engineers were in England working with their British colleagues on RJ500 design and development. The three Japanese companies rotated their engineering staffs through short-term assignments to expose the largest number possible to the British process of managing an engine development program. The tuition may have been high, but it could have been considered a doctoral program in systems management.

What did the British get out of it? Like the French and Americans, they probably saw Japan as a dangerous potential competitor. But they must have decided that this was an economical way to create a competitive new engine, one Rolls–Royce might not have been able to finance on its own; besides, Japan as a partner was perhaps less dangerous than Japan as a competitor—let alone Japan as a partner of the Americans.

The RJ500 program progressed slowly through the late 1970s. Early in 1981, Rolls–Royce told JAEL that the engine they had been designing and testing was almost certainly too small for the market as it was evolving; Boeing in particular wanted about 18,000 lb of thrust for the new 737. A redirection of the program would make the original decision look like a mistake—a difficult admission for any bureaucrat. To design a new engine now would cost large sums of money in addition to what had been spent on the RJ500; MITI would have to explain why the RJ500 effort had been wasted. It would not be easy to conceal that such a national program would have to make a major change of course. MITI and JAEL dithered. In March 1981, Boeing announced the launch of the 737-300 with the CFM56-3, whose thrust of 18,000 lb matched the specification of the 737-300 precisely. There was consternation in Japan. Imai, the president of JAEL, had maintained until the very end that Boeing would pick the RJ500. He resigned and soon after that retired from IHI as well.

Finally, MITI made the key decisions: the RJ500 program would not be killed but merely starved of funds until it expired quietly on its own, and a new company would be formed to develop an engine to compete with the CFM56. Rolls–Royce had been a capable and generous partner for JAEL, but both of them had guessed incorrectly on the RJ500. The Ministry of International Trade & Industry insisted that the new program have an American partner as well, because half of the world's market was in the United States, and it did not care whether the American partner was Pratt & Whitney or GE, either of which would give the program better access to the U.S. market.

GE offered JAEL a share of the CFM56 program despite SNECMA's grave reservations. The French, believing that Japanese involvement was probably against their interest, said that any Japanese participation would have to come out of GE's own share. The negotiator for the Japanese–British team was Rolls–Royce, which distrusted GE ever since GE objected to U.S. government loan guarantees to Lockheed after Rolls–Royce had gone into receivership. But there was a more fundamental reason why the Japanese were not ready to accept participation in the CFM56: a Japanese "national program" would enjoy national funding, but that required technology that could be represented as new. There, Japan could participate as an equal partner. The CFM56 program was a going concern. If any of the Japanese companies wanted to participate, it would be as junior partners to GE and SNECMA, and they would have to fund their own work themselves.

Instead, the Japanese offered GE a share of their new program. Because such a new engine would compete directly with the CFM56, GE could see no advantage in participating, and the Japanese and Rolls–Royce were not surprised. Pratt & Whitney, however, the other potential American partner, had no modern high-bypass turbofan of its own to replace the JT8D while it was continuing to make profitable sales of the refanned version. Pratt & Whitney, therefore, had something to gain and little to lose from joining the Japanese program. GE kept the dialogue with the Japanese going as long as it could, for almost a year, reasoning that the longer the negotiations, the longer the launch of a new engine would be delayed, and that such delays would benefit the CFM56. Finally the Japanese had enough and threw out the GE representatives. The Japanese engine companies, MITI, and Rolls–Royce selected Pratt & Whitney as their partner for the program.

They formed International Aero-Engines, Ltd. (IAE), with headquarters in Switzerland. Rolls–Royce and Pratt & Whitney each had shares of 30% in the programs. The three Japanese engine companies together had 23%, and the balance was divided between the German company MTU and the Italian FIAT. Soon after they started to design their new engine, the V2500, serious problems occurred in Rolls–Royce's development of the high-pressure compressor. To prevent a major disruption of the program, Pratt & Whitney took over leadership. As is not uncommon, development took longer than the IAE partners and their governments predicted. But the partners were capable and experienced. Despite the complex ownership, IAE and its owner companies ultimately succeeded in developing and certifying a high-performance turbofan engine. Airbus Industrie was happy to be able to offer airlines the V2500 as an alternative engine for the A320. But coming later, the V2500's market share inevitably remained smaller, and of course it was still in the relatively high-cost phase of the production learning curve. Airlines will not pay a higher price for the V2500 than they pay for the CFM56. It was a prescription for unprofitable sales.

When Airbus Industrie enlarged the A320 into the A321, GE and SNECMA offered a higher thrust model of the CFM56, as did IAE with the V2500. GE was shocked when its good customer Lufthansa selected the V2500 for its A321 but could only assume that no airline relishes being completely dependent on one supplier. Then Airbus Industrie needed an engine for its proposed four-engine long-range airliner, the A340. Again Lufthansa was expected to be a launch customer for the A340, for routes such as Munich–Johannesburg or Hamburg–Rio de Janeiro, which do not have enough traffic to justify a 747. GE and SNECMA offered yet another growth model of the CFM56, but Lufthansa had some doubts about their ability to enlarge the engine further. Airbus Industrie offered Lufthansa the A340 with a Superfan version of the V2500, and Lufthansa accepted. At that point IAE informed Lufthansa, with some embarrassment, that the "Superfan" was only a paper study and that IAE would not commit to certifying it for the A340. The technical director of Lufthansa, Reinhardt Abraham, was furious. It was he who had made the decision to pick the V2500 for the A321; it was he who had picked the V2500 Superfan for the A340. And now there was no Superfan. He reversed the engine selection for the A321, going back to the CFM56, and he accepted the bigger version of the CFM56 for the A340 also. It was a major loss for IAE.

Singapore Airlines also ordered the A340 and insisted on the same stringent specific range performance specifications that had brought the Douglas MD-11 to grief. Airbus Industrie was able to demonstrate that the A340 met its range specifications, no doubt to the relief of Lufthansa and CFMI.

Just as there is a market for versions of short-to-medium range airliners larger than the 737-300 and the A320, there is also a market for smaller versions. Many optimists in the European airframe and engine industries believe that this market is large enough to justify the development of a completely new airframe and engine. Fokker and Daimler Aerospace were for a time of this opinion; BMW formed an alliance with Rolls–Royce to develop an engine for such a new airliner. But there are many pessimists who believe that this market is limited, constrained from below by high-speed trains and existing aircraft such as Bombardier's Canadair RJ regional jet and from above by the McDonnell–Douglas MD-95, a shortened version of the MD-80 (after the McDonnell–Douglas merger with Boeing, this aircraft was rechristened the Boeing 717), and by versions of the 737 and A320. In fact, Douglas, Boeing, and Airbus Industrie could cut the prices of their airliners, which have been in production for some years, and make it impossible for a completely new airliner and engine to recover their development cost.

Such a debate began to take place during the recession of the early 1990s, which depressed the growth of air traffic in Europe and was encouraging the consolidation of the airline industry in the European Union. Daimler Aerospace has announced that it will no longer support Fokker, its subsidiary, with working capital and it is likely that Fokker will be liquidated. Daimler has also expressed chagrin at the unattractive profit performance of its engine subsidiary MTU.

Airbus Industrie has always believed that it must offer a broad range of related aircraft to compete effectively with Boeing. It is not surprising, therefore, that Airbus Industrie is offering a shortened version of the A320, the A319, with CFM56 or V2500 engines. The design may not be as efficient as a completely new aircraft and engine, but there are great savings to the airlines in a commonality of airframe systems and engines, and the possibility of a lower price because of the learning-curve effect on production costs. From the airlines' point of view, such an aircraft can be more economical than a new design. That has always been Boeing's marketing strategy.

In the meantime, the three Japanese engine companies had established themselves as partners in the V2500, a competitor for good or ill to the CFM56 on the world market. All three of them participated in the subsequent development of large engines: Mitsubishi with Pratt & Whitney, Kawasaki with Rolls–Royce, and IHI with Rolls–Royce and GE. And when GE launched the CF34-8 engine for 80–120 seat airliners, IHI and Kawasaki both signed up to participate, having concluded that it was a better strategy for the moment to be junior partners to all big three engine companies than to attempt to take the lead themselves.

Chapter 15

Advanced Technology Programs

IT TAKES astute (or lucky) product planning to create superior products, and engineering programs to create the limiting technology. American industry, in general, spends about 3% of its revenues on research and development. High-technology companies spend more, the semiconductor industry as much as 10%. Aircraft engine and perhaps pharmaceutical companies are the biggest investors, with as much as 15% of revenues going into research and development. Why are there so few aircraft engine manufacturers? One of the major reasons is the huge level of investment required. The design and certification of a new commercial engine requires something in the neighborhood of $2 billion, spent over some five or six years. But before that process can even begin, the enterprise has to have at its disposal the basic and applied technologies needed for the design of a specific engine. For a company like GE or Pratt & Whitney, the 15% of revenues spent on research and development breaks down roughly as follows:

- Initially, 3% is spent on basic research and development, for example, the development of new materials, new computational techniques, the basic physical chemistry of combustion, aerodynamic and structural analysis, new concepts of manufacturing and inspection such as computer-aided tomography.
- Another 4.5% is spent on applied research and development or enabling technology for the engineering, manufacturing, and quality control processes.

- Finally, 7.5% is spent on product engineering, design, testing, repair technology, and correction of problems.

For a company with annual sales of $5 billion, this means spending $750 million/year on technology. Whether this makes sense can be measured by a single criterion: whether the advanced technology required is available and experimentally validated at the time needed for a new product. Without such validated technology, the next generation of products will not be available at the right time or at a price the market is willing to pay.

Through the 1960s, the U.S. DOD funded a substantial portion of aircraft engine research and development, as governments have done in other countries. Since then direct U.S. government funding has declined, but the engine companies have held steady the portion of their total revenues devoted to R&D. Saving money here goes to the bottom line at once but may short-change the future, as in the case of Wright Aeronautical Division. In contrast, spending money now on R&D may enable the development of a new product in the future, but if R&D does not lead to a product wanted by the market, the money will also have been wasted to some extent. Three examples from GE Aircraft Engines illustrate the choices that have to be made.

Quiet Clean Short-Haul Experimental Engine

In the early 1970s, NASA launched a program for a "Quiet Clean Short-Haul Experimental Engine," or "Quick See" (QCSEE). The objective was a powerplant for a new generation of commuter airliners for intercity traffic in the northeast corridor between Boston, New York, Philadelphia, and Washington, D.C. New airports were under consideration, to be built close to city centers. Aircraft had to be quiet and capable of steep climb after takeoff and steep descent before landing to minimize the nuisance noise "footprint" around airports: the smaller the noise footprint, the fewer the angry voters living near the airport.

Airports have dealt with the nuisance noise problem in several ways. At Kennedy Airport in New York and Heathrow Airport in

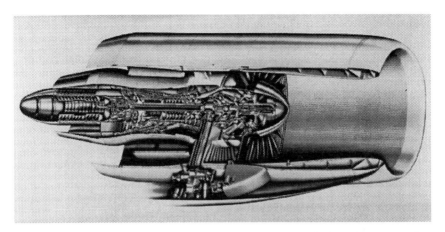

Fig. 7 QCSEE gear-driven very high bypass ratio turbofan. (GE negative)

London, for example, pilots are forced to make tight turns shortly after takeoff so that the path of their climb avoids residential areas as much as possible. Pilots prefer not to make such turns at low altitudes. Some airports limit access by older, noisier aircraft, to which airlines have responded by using quieter aircraft. Aircraft with high-bypass turbofan engines typically meet "Stage 2" noise levels, while earlier airliners such as 707s, 727s, and DC-8s meet only the "Stage 1" noise levels and may not be used on international flights after 1998. Other airports pay for the soundproofing of houses most affected by airport noise and occasionally even buy such houses so that their former owners can move elsewhere, out of the noise "footprint."

Quick See was aimed at the first two options: a high bypass ratio for an inherently quieter engine, and power adequate for steep ascents and descents with full aircraft stability and control. Two contracts were awarded by NASA, one to Hamilton Standard (a sister company of Pratt & Whitney in United Technologies Corporation) for an advanced variable pitch propeller and gearbox, and a second to GE, for a new turbofan engine (see Figure 7). GE took this as a major technology challenge. A. P. Adamson, one of its senior designers, led the project. The technical facts were not clear, and GE decided to design and build *two* versions of QCSEE, one with a fan with fixed-pitch blades, the second with variable-pitch blades. Both would use the same core engine to drive the fan, the core of the military F101, which was readily available, had high specific performance, and had been designed from the beginning with the airflow and pressure ratio that would allow GE to use it as a building block for the development of other turbofan engines. (GE used this core engine for the F101 engine for the B-1 and B-2; subsequently for the F110 engine in the F-14 and F-16; and for the CFM56 commercial engine used in the Boeing 737, the Airbus Industrie A320, and the U.S. Air Force KC-135).

The variable-pitch fan presented a major technical challenge and GE developed two separate and different pitch change mechanisms, one designed by GE itself, the other by Hamilton Standard, the propeller specialists. Both designs worked well enough for a demonstrator engine but would have required considerable refinement for a production engine. They had very high bypass ratios and were shrouded by casings like conventional turbofan engines. To allow the tips of the big fan blades to run at subsonic speed to minimize noise, the fans needed a lower rotating speed and were driven through reduction gears from the low-pressure turbine (GE selected Curtiss–Wright to design the reduction gearbox). GE built and tested both versions of QCSEE, and both demonstrated performance that showed that NASA's objectives could be met. But neither city-center airports nor airliners capable of short takeoff and landing were ever developed, so that QCSEE remained a technology demonstration without a direct product application. In consequence, GE never did the iterative development of the fan, turbine, and nacelle aerodynamics that a good production engine would have required.

Nevertheless, GE and NASA learned some valuable lessons from QCSEE, primarily about the design of large gear-driven fans and the aerodynamic drag of such large-diameter fans and their nacelles, which turned out to be less than had been predicted. GE was later able to apply the technology of the large diameter fan to the GE90 engine for the Boeing 777. GE also developed valuable information for the design and fabrication of composite structures—the fan blades, the fan casing, the reduction gearbox casing, the front frame, and the exhaust nozzle were all composites of glass cloth, plastic resins, and metal reinforcements. GE decided on composites because it needed quick delivery and modest tooling cost for such an experimental engine program. They served the QCSEE purpose admirably and could have been developed into production parts for a definitive engine, with one exception: the composite fan blades could not withstand the bird strikes specified by the FAA for certification (the "Hyfil" composite fan blades of the Rolls–Royce RB.211-22 also failed to survive hailstones and bird strikes, contributing to that company's going into receivership in early 1971). GE made good use of this composite information in the subsequent unducted fan technology program, and finally managed to solve the bird strike problem there and on the commercial GE90 engine.

Energy Efficient Engine

Following the "oil shock" after the 1973 Arab–Israeli war, NASA decided to investigate the technology required for a radical improvement of the fuel efficiency of commercial aircraft engines, the so-called energy-efficient engine ("E Cubed"). GE's preliminary design group in Lynn studied various engine cycles that could yield better fuel efficiency, seeking to find the optimum balance between bypass ratio, overall engine pressure ratio, turbine inlet temperature, and mixed or separated flow of the fan and core engine gas streams. GE wanted to determine where new component technologies would be required if cycle studies, which are cheap and hypothetical, were to be turned into a real machine. The studies showed big gains from a new compressor. GE had been studying compressor technology for some years to optimize efficiency, weight, and cost. Calculations showed that the E Cubed would require a compressor with a pressure ratio of 23 in only 10 stages (diesel engines, considered very fuel efficient, have pressure ratios beginning at around 15:1). Such unprecedented performance required novel aerodynamic design for the compressor blades, unusually high rotation speeds, and new materials to withstand the aerodynamic loads, to keep the blades from untwisting, and to retain them in the rotor under the centrifugal loads.

Fig. 8 M. C. Hemsworth with the E Cubed demonstrator engine. The bunch of pipes sticking out of the turbine exhaust is part of the sound suppressor. (GE negative)

In 1975, NASA awarded E Cubed contracts to both GE and Pratt & Whitney. Each company received $90 million from NASA and had to invest $10 million of its own, with zero profit, to demonstrate the technology required to achieve a specific fuel consumption 12% lower than that of its current production airline engine—the JT9D for Pratt & Whitney and the CF6-50 for GE. Pratt & Whitney focused on component technology and developed a very advanced single-stage high-pressure turbine, running at a reduced hot gas Mach number to improve its efficiency and reduce losses induced in the low-pressure turbine (GE was also using a single stage high-pressure turbine in the F101 engine, whose core was in QCSEE and the CFM56 commercial engines). The GE project manager, M. C. Hemsworth, attacked the E Cubed program in a more comprehensive way, looking for the best new high-pressure compressor, fan, and low-pressure turbine that could be integrated to a complete engine system (Figure 8). GE wanted to achieve an improved system rather than results on isolated components, and set itself a schedule to that end, with resources to match.

The low-pressure turbine proved to be a difficult aerodynamic challenge. GE made a number of tests of models and full-scale parts of the first two turbine stages and, by late 1979, completed all of the component tests, including tests of the remarkable new high-pressure compressor and a novel "double annular" combustor that exhibited much lower noxious exhaust emissions, including emissions of nitrous oxide. In a conventional combustor, the normal rule was that the higher the combustion temperature the better the fuel efficiency, but the worse the generation of nitrous oxide. With the component tests, GE demonstrated that it had achieved all of NASA's *component* design objectives. GE then assembled the components into a complete system, taking care not to call it an engine but a "component technology demonstrator." GE's E Cubed demonstrated a specific fuel consumption 13% lower than that of the CF6-50. GE exceeded the original challenge of a 12% reduction and NASA saw the test as a validation of its leadership. GE saw it the same way for its own technology.

GE immediately applied the lessons of E Cubed technology to its product design for the CF6-80E engine, the latest model for the Airbus Industrie A330, and to the new GE90 for the Boeing 777, which uses the complete E Cubed high-pressure compressor more or less unchanged.

The Unducted Fan

In 1977, when jet fuel prices to the airlines had risen from $0.30 to more than $1.00/gal, Boeing was developing the 757 as a 175-seat successor to its 727. Rolls–Royce and GE both proposed engines using existing cores from their trijet engines. Pratt & Whitney's JT9D core was too big; so Pratt & Whitney proposed a completely new engine, the PW2037. Pratt & Whitney designed a very advanced engine and promised the airlines a fuel consumption 8% lower than that of any competitive engine on the market; if Pratt & Whitney was unable to deliver, it promised to pay the airlines the difference in operating cost. Delta Airlines and other American carriers accepted this riskless proposal. GE, concluding that this was a competition that it could not win financially, stopped development of its CF6-32 engine. Rolls–Royce continued with its development of the RB.211-535 and, in time, won about half of the Boeing 757 market.

The year 1980 brought another oil crisis. The general manager of GE's engineering division pointed to a White House study that predicted fuel prices of $4.00/gal and challenged Arthur Adamson to find a solution to such a fuel price, a fuel reduction radically greater than the 12% specified by NASA for E Cubed five years earlier. Adamson had led the QCSEE project and had been the chief designer for the CF6-50 engine. His optimization studies led him quickly to prop fans. Fuel efficiency and noise limits require low tip speeds for the fan blades and, therefore, slow rotation speeds for the fans, which Adamson had accomplished in QCSEE with a speed-reduction gearbox between the low-pressure turbine and the fan rotor. Airlines avoid such gearboxes when they can because of their complexity and high maintenance requirements. For an engine of the size needed by a 150-seat airliner, the gearbox would have to transmit more than 20,000 hp, a major engineering challenge.

One of Adamson's engineers invented a counter-rotating turbine to drive two counter-rotating fan stages, a concept GE called the "unducted fan," or UDF. One stage of the fan blades is driven by the low-pressure turbine rotor, the other stage by the low-pressure turbine *stator*, rotating in the opposite direction. Each can turn at a relatively low absolute physical speed, less than 1500 rpm, producing low tip speeds for the fan blades, good fan efficiency, and acceptable noise levels. Counter-rotation of the second fan stage also straightens out some of the swirls created by the first stage and improves fan efficiency by some 7%. Between the low-pressure rotor and stator, the relative speed is twice the 1500 rpm. Because turbine work is proportional to the square of relative speed, doubling the relative speed of rotation allows fewer turbine stages for a given power output.

A rotating stator is a contradiction in terms. The mechanical design and assembly procedures for the UDF challenge the imagination. Everyone but GE laughed at the idea. GE encouraged Adamson to proceed and set aside $100 million for the project, which would use the high-pressure core of the F404, a smaller version of the F101, to drive the novel low-pressure turbine of the UDF. Intrigued by the idea, NASA made another $10 million available and stimulated Pratt & Whitney to study novel prop fans, in Pratt & Whitney's case gear-driven (see Figure 9).

The UDF fan blades themselves are about 5 ft long, plastic composite airfoils built around a titanium spar. Because of the composite material, each blade weighed only about 10 lb. GE demonstrated that the blades, rotating at less than 1500 rpm, would not penetrate the pressure cabin of an airliner if they came free from the hub, also that the blades could withstand the specified FAA bird strike and continue safe engine operation. The QCSEE lessons had been put to good use.

Fig. 9 Unducted fan running on outdoor test stand. The engine air inlet is at the right, the exhaust at the left. The two stages of counter-rotating fan blades can be seen to the right of the long tail cone. (GE negative)

The industry scoffed at the notion of the UDF. Neither passengers nor the airlines would ever accept a propfan after having been flown in jets. The aviation magazine *Flight International* derided the UDF as "a bunch of whirling bananas." GE decided on a bold gesture and built four UDF engines with F404 cores. After ground testing demonstrated that the concept would indeed work and established a baseline for engine performance, GE and Boeing installed a single UDF on one side of a Boeing 727, replacing the normal JT8D engine. Then GE and Douglas did the same on a DC-9. Groups of airline executives, NASA and FAA officials, and aviation journalists were taken for demonstration rides. The JT8D engines were throttled back and the airliners cruised on their UDF engines.

The results were dramatic. The UDF noise was quite different from that of a conventional jet engine, more like the sustained hum of a sewing machine, in the words of a sympathetic observer, and not unpleasant, according to the few selected airline passengers who were exposed to it. Most important, the specific fuel consumption was almost 40% lower than that of the old JT8D engines installed in the 727 and DC-9, and at least 30% lower than that of the best turbofan engines then flying! This was a staggering improvement in fuel economy.

GE and Douglas took the DC-9 to the Farnborough international air show near London, to amaze their friends and to scare their competitors. The flight demonstration was a *tour de force*. The UDF itself had been a technology demonstrator only but, having shown that the concept was valid, GE could now provide preliminary performance data for a definitive engine, the GE36, to the airlines and airframe companies.

GE's partner, SNECMA, on the CFM56, was horrified. Its forecasts of company revenue and income depended on the success of the CFM56 in the Boeing 737 and Airbus Industrie A320. An advanced propfan like the GE36 could cut that short, could cannibalize sales of the CFM56, and might even threaten the future of the A320, which had just been launched. Nevertheless, SNECMA, understanding the tyranny of the market, insisted that it be given a substantial role in the design and development of the GE36 and its production to compensate for any adverse effects on the CFM56 program. GE agreed, and SNECMA took responsibility for the design of the GE36 high-pressure compressor, a much more critical role than SNECMA had played on the CFM56.

Then the price of Saudi light crude oil stabilized at about $20 a barrel, and jet fuel dropped to $0.50/gal, a level much less critical to the operating costs of existing aircraft. The airlines weighed that against the cost of investing in an entirely new generation of short-range airliners designed to make effective use of the UDF and concluded that this was no longer so urgent. The GE36 program was shelved. No doubt Airbus Industrie and SNECMA breathed a little easier.

With the encouragement of NASA and to defend against the UDF, Pratt & Whitney had started development of an advanced propfan of 25,000-lb thrust, comparable to that of the UDF, its blades driven by a large gearbox designed by FIAT. Such an engine would have been mounted like any conventional turboprop, perhaps more conveniently than the UDF. But like the UDF, this program was shut down when oil prices dropped to the level of $20 a barrel.

Adamson viewed the UDF program as a failure. So much money spent, so much creative engineering, and no engine to show for it. In addition, GE did not solve the problem of mounting such an engine on the wing as readily as on the sides of the rear fuselage. Yet, if the UDF did nothing else, it established GE as a daring and successful originator of advanced aircraft engine concepts, a technical leader of the industry. That is not a small advantage when one is marketing to the airlines.

CHAPTER 16

Developing the Chinese Market

ACCORDING to Senator Joe McCarthy, there were Communists in the U.S. State Department who "lost" China to Mao Zedong in 1949. In Korea, the Chinese People's Liberation Army entered the war on the North Korean side in 1950; Chinese communist influence was seen behind the civil uprising in Indonesia in 1965; China was the sinister force behind the North Vietnamese and the Vietcong. For American aerospace companies, China, long a Soviet ally, was simply enemy territory, off limits and unexplored.

In 1972, U.S. President Nixon and U.S. Secretary of State Henry Kissinger, in an unexpected reversal of policy, reestablished diplomatic relations between the United States and the People's Republic of China. This may have been to cover the northern flank of the U.S. withdrawal from Vietnam, perhaps to put pressure on the former Soviet Union by taking advantage of Soviet–Chinese rivalry at their border in Manchuria, or it may have been a simple acknowledgment of reality. Whatever the motives, the U.S. government encouraged American firms to explore commercial opportunities. GE, like most Americans at the time, knew little about this big socialist command economy.

The Civil Air Administration of China (CAAC) not only owned the airline that flew domestic and international routes, but was also the operator of hundreds of agricultural and other aircraft, the air traffic control system, and the Chinese safety certification agency. The CAAC reported directly to the Chinese State Council, the cabinet of the PRC. Its airline used Russian-designed airliners, some of them built under license in Chinese factories. These factories were owned by another ministry not part of CAAC. To supplement its Russian equipment, CAAC bought some foreign equipment, including 13 used DeHavilland Trident airliners from Pakistan International Airlines for its internal routes. The Trident is powered by three Rolls–Royce Spey engines. After the resumption of relations with the United States, CAAC ordered 10 Boeing 707 airliners with Pratt & Whitney JT3D engines for its international routes, plus one spare engine for each installed in the aircraft, a much higher proportion of spare engines than any other airline. Perhaps the Chinese were protecting themselves against a repetition of their experience with spare parts supplies from the former Soviet Union. In any case, within a couple of years CAAC found itself with modern airliners using Rolls–Royce and Pratt & Whitney engines but no engines from GE. The CAAC had not set out to buy Rolls–Royce or Pratt & Whitney engines as such. They simply came with the airliners the Chinese bought. In late 1978, GE gave me the responsibility for developing the market for GE engines in China.

GE's Marine & Industrial Engine Division began discussions with the Chinese government in mid-1978, sending brochures to Beijing through a liaison office in Hong Kong and then holding a technical seminar in China in the fall of 1978. An office in Hong Kong approaches the Chinese Ministry of Foreign Economic Relations & Trade (MOFERT) and asks it to distribute technical literature to potentially interested ministries and agencies. After some time, if MOFERT has found enough interest, it issues an invitation to the foreign company to hold a technical seminar at which the interested Chinese can ask questions and learn more about the product.

GE may have faced some political obstacles. When the Communists took power in 1949, they seized and nationalized all privately owned businesses, among them a light bulb factory in Shanghai owned in part by GE. GE was one of many expropriated American companies that filed legal claims against the government of the PRC, causing some threat of seizure by legal writ when Chinese ships or aircraft landed in the United States. After the Korean War, GE also built up a big business presence in Taiwan, the Republic of

China (ROC), which the PRC regarded as a renegade province, threatening to reunite it with China by force of arms. On its side, the government of the ROC had some pretensions to being the legal government of mainland China. Companies like GE that conducted business in Taiwan and had old property claims against the Communists faced a headwind trying to do business on the mainland. Or so we thought.

Selling engines to an airline requires a detailed understanding of the airline's technical objectives, route structure, plans for future passenger and freight capacity, maintenance plans, and access to financing. Aircraft and engines are very expensive and must stay in service for many years. Their selection is a matter of long-term importance to the airline and to its owners, particularly when the owner is a government.

Airlines expect competitive performance and a competitive price from any engine proposal. Sometimes they must be led through price calculations over the entire service life of an engine so that they do not become fixated on the initial acquisition cost alone. They expect good product support and a supply of spare parts at acceptable prices; high reliability; long parts life; and easy inspection and maintenance. Their governments may have other requirements as well, such as industrial development or offset purchases by the seller. All of the competitors normally offer to meet whatever requirements the airline sets. Clear-cut technical superiority or a significant advantage in reliability gives a big edge, but in the end another important ingredient is the airline's confidence that the seller's word is good, it will deliver precisely what has been promised, and it will be there to help when there are problems. There will always be problems.

The most convincing evidence is demonstrated performance. But, in addition, the seller has to cultivate close personal relationships with the people in an airline who evaluate competing proposals, and those who make the final decision and understand what is important to the customer. The typical airline sales campaign begins several years before the official order is signed by a purchasing agent.

GE Aircraft Engines assigned marketing responsibility for each potential customer airline to an account manager in the Airline Marketing Division, supported by marketing people located as close to the customer as makes business sense. GE technical people were stationed at the aircraft factories of Boeing and Airbus Industrie and Douglas. There are GE field offices in concentrated major markets—France and Germany, Britain and Japan, Brazil and Singapore, and the Middle East. The largest single market was and is in the United States and Canada, and this is served from the Evendale, Ohio, headquarters of GE Aircraft Engines, supported by a small team in Canada.

Customers like to believe that the marketing people with whom they deal can make decisions and commitments on behalf of the seller's headquarters. The Airline Marketing Division's account managers were on the road more than half the time, visiting customers in support of sales campaigns or merely keeping in touch for the next opportunity. The field office marketing people also rode their circuit constantly to keep customers informed of what was happening in the Evendale headquarters and to keep Evendale informed of what was going on in the market.

In the major industrial countries, and particularly where there was an established successful relationship with a customer, the account managers in the Airline Marketing Division carried the major marketing load. But there are many markets where English is not the vernacular language; where society, culture, laws, and customs differ from those in the United States; and where GE salesmen are as exotic to the customer as his country might be to them. There the Airline Marketing Division relied more heavily on the people in the field offices, whose responsibility it is to bridge the gap between the customer and the factory marketing people. In some countries, where it had no prior experience, GE engaged consultants who could advise it on strategy. For example, at one time a field marketing manager stationed in Rome visited airlines in West and Central Africa every two months, depending on a former diplomat in the Organization of African Unity for advice and information. As GE established effective working relationships with the airlines, these responsibilities were taken over by the designated Airline Marketing Division account manager.

With finite resources, priorities must be set. The Airline Marketing Division concentrated on North America, Western Europe, East Asia, particularly Japan, and the Middle East and South America. That was where the action was, the big airline accounts in the countries with big populations and the financial resources to invest in airlines and air travel.

During World War II, Gerhard Neumann had served as an aviation mechanic and a technical intelligence specialist in Claire Chennault's American Volunteer Group, the famous "Flying Tigers." GE had no contact with the Chinese aviation community since 1947, when Neumann joined Chennault's fledgling airline Civil Air Transport in Taiwan as chief engineer. As far as the mainland was concerned, Taiwan was the wrong side. In the years after the war, GE had also built up an active trade in Taiwan in power generation, plastics, and many other products. Any relations with the PRC would have to be built from the ground up. Finding one's way in a socialist economy such as China and later Russia was a new problem for GE, different from the conventional airline sales campaign.

An Old China Hand

An incident 10 years earlier, in 1968, illustrates some of the differences between East Asian business culture and commercial practice and those in the United States. I had been in Japan to negotiate a license agreement with Ishikawajima-Harima Heavy Industries, Ltd. We had worked together on other programs for eight or nine years before that, and quickly reached agreement on a collaboration that has continued to this day. We celebrated the signing of the agreement with a small dinner, four Americans from GE and the Japanese managing director of IHI. He had been born in Brooklyn and his parents returned to Japan when he was two years old. None of the four of us from GE had been born in America—our resident manager was born in Germany, our contract specialist in Brazil, our lawyer in England, and I in Austria. Dr. Imai wore a Swiss watch. The four of us wore Japanese watches. Soon after the dinner, we received a message. There was a problem with GE's sales representative in Hong Kong. Because Hong Kong is closer to Tokyo than to Cincinnati, we were to go there and fix whatever was wrong. The lawyer and I decided to fly to Hong Kong, leaving our contract specialist in Tokyo to tidy up the license agreement.

GE Aircraft Engine's sales representative in Hong Kong was an American, Pat Paterson, whom GE had retained several years before to solicit orders for GE helicopter engines and spare parts from customers in Hong Kong and Thailand. Paterson was then in his late sixties. He had come to China in the late 1930s to sell Brewster fighters to various Chinese war lords and ultimately to the Nationalists, had been interned by the Japanese during the war, and had stayed in Hong Kong afterward as a sales representative for American and European aviation companies. Paterson was eminently qualified for the title "Old China Hand." We met him in the bar of the Overseas Press Club in Hong Kong. He was having drinks with an old friend, who had been the Hong Kong manager of National City Bank when the Japanese arrived in 1942 and had the foresight to take a lot of the bank's cash with him into internment with all of the other American civilians. He and Paterson had been friends before and during internment. The banker was able to make friendly loans to Paterson and many other prisoners to tide them over during the hardships of prison. None of the loans were defaulted, he told us with pride. Now the banker was retired and was helping Paterson in some capacity to keep from being bored. He wore a bright red nubby silk suit, so that boredom probably was not too serious a problem.

Over drinks, Paterson made clear that he was very angry with the GE helicopter engine product support people. They were idiots, he said, who did not understand the customer, and they were totally unresponsive to the market. As an example he offered his current predicament. One of the customers in Paterson's sales territory, the Thai Provincial Police, operated a single Sikorsky S-62 helicopter with a GE T58 engine. Paterson sent a telex to the GE factory requesting an urgent quotation for an instrumented engine test cell "just like the one you have sold to the Royal Malaysian Air Force." Our people replied to his telex by air mail, stating that the Thais did not need so elaborate an installation for one helicopter and proposed a portable test cell to be mounted on a truck bed. The letter took 10 days to reach Hong Kong.

Paterson was furious. He called up the factory and explained that when he requested a quotation by telex, that was a good signal that the response should come faster than by mail. How much did a portable cell sell for? What was the price of the instrumented test cell GE had sold to the Malaysians in Kuala Lumpur? The factory gave him the answers. With those numbers, he prepared his own proposal on GE stationery that he just happened to have in his office and typed up a *pro forma* invoice purportedly issued by GE. He gave these to his customer in Bangkok. At that point, GE asked our lawyer and me to pop over to Hong Kong.

It turned out that Paterson was right: GE did not understand the customer and was not being responsive to the market. The Thai Provincial Police definitely wanted a test cell much more elaborate than the one GE considered appropriate—the more expensive the better. The uncle of the lieutenant in charge of procurement was the contractor who would build the cell and furnish its equipment: the bigger the job, the more benefit. Similar considerations applied to Thai orders for spare parts.

This was before the days of the Foreign Corrupt Practices Act. In any case, there was no bribe involved here, not by GE, and not, I believe, by Paterson either. We were dealing with a customer who had decided that he knew better than GE what he wanted, in part to enrich his relatives and perhaps himself. This did not appear to be an unusual or reprehensible notion in Asia. As long as neither we nor our sales representative were committing a crime, as long as we were neither misrepresenting our product nor causing the customer physical harm or danger, did we have an obligation to insist that he buy only what we recommended rather than what he wanted? I recommended that we sell the Thais the big test cell. By the time our people in Lynn had come around, another contractor got the order. Paterson was not happy about that, but somewhat mollified that now GE at least understood what was going on.

Learning the Ropes

GE set up a China liaison office in Hong Kong, managed by Ed Naylor, who certainly looked like an Old China Hand to those of us new to the game. In the nature of GE reorganizations, Naylor disappeared fairly quickly and was followed by Dick Kask, born in Shanghai to American missionary parents. Kask was also moved quickly to another assignment. Finding the right person for the Chinese market was obviously causing some problems, problems GE Aircraft Engines would also encounter. Individual GE businesses began to explore the market for capital goods involved with the infrastructure of the Chinese economy: steam turbines and electric generators for powerplants, locomotives for freight and passenger trains, and electric wheels for ore trucks used in coal mines. Our Marine & Industrial Engine Division wanted to market a lightweight gas turbine for ship propulsion. And our Commercial Engine Operation wanted to sell airliner engines to CAAC.

The marine engine people chose Robert Olson to go to Hong Kong to make contact with the Chinese navy and China State Shipbuilding Corporation. Olson was a bachelor who had at one point spent several years serving in the Peace Corps in West Africa. It was not difficult for GE to persuade him to live in a hotel room in Hong Kong and learn how to make frequent two-week trips to Beijing. Olson was very enthusiastic about his new assignment and took up intensive study of the Mandarin language and Chinese culture.

The Chinese language is an enormous practical obstacle for foreigners in China. Very soon, therefore, GE sent Olson an assistant who was cut from a different mold. Walter Chang had been born in 1940 in what was then Peking. His parents fled to Taiwan with him in 1948; there, in due course, Chang earned an engineering degree from the National Taipei Technical University and then served in the ROC air force. He came to the United States and earned a Bachelor's degree from the University of Massachusetts and a Master's degree in aeronautical engineering from Worcester Polytechnic. After graduation he went to work for GE Aircraft Engines in Lynn as a compressor designer. In 1974, when GE was cutting back staff, he took a leave of absence to become the industrial commissioner for the city of Fall River, Massachusetts. Chang visited his parents in Taiwan, and there was approached by a couple of garment manufacturers who wanted to enter the American market; Chang persuaded them to set up factories in one of the many empty mill buildings in Fall River. In 1975, GE invited him to come back to work. Chang first took an assignment in evaluation engineering to broaden his experience but from the beginning had his eye on a career in marketing. It was clear that he had unusual energy and the local knowledge GE would need to develop the aircraft engine market in China, and GE quickly assigned him to support the airline salesmen assigned to the accounts in Taiwan and China. Chang worked well with some of them but was frustrated by the Airline Marketing Division's insistence on controlling all aspects of such activities and its reluctance to accept help from others outside the division. After a trial period, the GE personnel chief suggested that I use Chang in my new Chinese market development activity. That would give him more scope to use his talents. I asked Chang to help Olson in the marine engine marketing activities.

Chang quickly took over in Hong Kong. Discussions with customers began to be held mostly in Mandarin—comfortable for them because there was no need for translation. Soon he no longer bothered to tell Olson everything that had gone on and even questioned on occasion whether Olson could add any value to the transaction. Chang was not always an easy man to work with, but he was energetic and effective with customers. Perhaps it was too much for Olson, who soon returned to the United States. He took early retirement from GE, after fitfully writing an account of his experiences. I asked him for a copy. He gave it to me with reluctance and had obviously taken great pains to write with tact about difficult personal relationships.

The Peking Hotel

I went to China for the first time in April 1979. We had been invited to Beijing to give a technical seminar on our commercial engines. Our delegation was led by Ed Woll, the retired vice president of engineering. It was a shrewd choice: Woll had the rank and reputation and the time to devote to what might turn out to be a trip down a blind alley. As cultural preparation, we brought an anthropologist and an historian to our headquarters in Cincinnati to give us a quick course on China. One of them drew a circle and around the rim wrote the names of continents—Asia, Africa, Europe, America. In the middle he wrote "China," a graphic representation why the Chinese name *Zhonghua* means the Kingdom at the Center of the

World, or "Middle Kingdom" as it used to be called. One of our people snorted derisively. The professor erased the word "China" and wrote "USA." We all smiled—seeing ourselves as the navel of the world is a characteristic Americans share with the Chinese.

Before we could start with our technical seminar in Beijing, there had to be a meeting at the Third Ministry of Machine Building, which was responsible for China's aviation industry, but not, we were soon to learn, for the airline. The protocol was quite rigid. For meetings with the higher ranks of Chinese, the conference rooms were furnished with two rows of heavy upholstered arm chairs, with crocheted or cotton dust covers over the arms and back. There were low tables between the rows of chairs holding ash trays and covered porcelain tea cups or soft drinks. One chair was placed at the end, facing the two rows. This was where the host sat, the minister or chairman. Chairman is a universal Chinese title, so Ed Woll as the senior member of the GE delegation became our chairman. The Minister of the Third Ministry of Machine Building sat at the end, an interpreter behind his left ear. Chairman Ed Woll sat on the minister's immediate right, the rest of us arranged down the row on his right. The Chinese sat on the other side, on the minister's left. There was no particular order to their seating: some senior people sat in the middle, some at the end. In such a formal meeting only the minister and our chairman would speak, the minister of course first with a fulsome welcome. The rest of us looked as grave as we knew how and studied the Chinese.

Our Chinese experts warned us that ministers and other senior Chinese officials often pause for quite a while between thoughts, even beyond the time for translation. We should wait for at least 30 seconds of silence before we assumed that the minister had in fact finished and jumped in with our replies. Waiting was a difficult concept for Ed Woll, whose quick mind was matched by his rapid fire talk. Woll had great trouble waiting for the minister to finish, and then he usually forgot to wait for the interpreter to catch up with the translation of his own words. The minister did not seem perturbed. He simply continued with his own speech whenever he was ready until he had said everything on his mind. Then he would sip tea while Ed responded. After a few funny flubs, Ed caught the rhythm of such formal dialogue. The process did not give much scope to any informal expression of ideas.

Such informality did emerge when we finally got to the technical discussions, which ran at a different pace. Our people addressed the subjects in which they were expert with the same informality they used at home. So did the Chinese, many of whom understood some English. Whenever the interpreter stumbled over a technical phrase, everyone pitched in to find the exact translation.

This was in the spring of 1979. The walls facing the street were covered with hand-printed posters, paper sheets 2 ft by 3 ft covered with writing, the famous Democracy Wall where anyone with a strong political idea was prepared to expose it to public debate. Groups of citizens studied the posters and discussed them with animation. Like most westerners, I had no idea of the significance of what we were seeing.

Once a man abraded by a hard life and with only a few teeth left came lurching toward us as we walked and said something, not very friendly from the looks of it. Other passersby surrounded him and pulled him away, apologizing to us for the inconvenience.

Communist Party reaction to the criticisms expressed in many of the posters came quickly. Before our second visit two months later, Chairman Deng ordered a crackdown on political expression. The posters were torn down, and the authors of the most outspoken thrown into jail. But at that time, as we walked, several young people approached us and asked if they could practice their English. They asked us what we were doing in China. They said, "America? Ah—The White House must be very beautiful." Was there more to that than stilted phrases from English lessons? We were careful to stay away from any remark that might be patronizing or provocative, which left us only platitudes. In contrast, on his first visit to Beijing since leaving in 1949 as a small refugee with his parents, Walter Chang immediately engaged in animated political discussions with the floor people in the hotel, people in shops, or people on the street. Reckless, perhaps, but a lot more interesting than what we had done.

The Ministry had assigned an interpreter to us, a young woman about 25 years old who spoke perfect English, which she learned at the Shanghai Foreign Language Institute. She knew less about Chinese history than we did because she was of the generation in high school or at the university during the Cultural Revolution, when education essentially came to a halt. The schools were emptied of their students, who organized themselves into regiments to terrorize their teachers and parents. They fanned out all over China to carry the message that all figures of authority other than the party and the army were corrupt. Exercising authority over one's elders must be heady stuff for rebellious teenagers. They assisted Chairman Mao's purpose of preventing the emergence of any alternate power structure. But they certainly did not go to class.

So there was, in 1979, a generation of young people who had skipped the completion of secondary education completely. They were unprepared for the university and too old to start again in competition with the next age group of youngsters. For a number of years, these young people had been used to exercising brutal public authority. Now they had no authority and no training for any career. The most fortunate, or the best connected, our interpreter among them, found work in the hotels or with government agencies and received at least some training.

Missteps by the Chinese Industry

Before we could enter the Chinese market effectively, we had to understand the structure of the industry. On the Russian pattern, research was done at research institutes, for the aviation industry, for example, by the Northwest Polytechnical Institute in Xi'an. Design and development were done by design bureaus, usually clustered around research institutes. In fact, all of the aircraft and engine designs we saw in China had come from other countries. The Third Ministry of Machine Building fanned out the production work to factories all around China, whose only technical contribution was production engineering of the product design they had been handed.

The engine factory in Xi'an built Spey engines under a Rolls–Royce license with the help of a large group of English technicians, for installation in a Chinese bomber designed in Xi'an. Another engine factory in Harbin in Manchuria built Wright Aeronautical Twin

Cyclone engines under a Soviet license, down to nameplates in Cyrillic, and was experimenting with a Soviet small turboprop engine. A factory in Chengdu, in Yunnan, assembled turboshaft engines for helicopters under a French license and made parts for small Canadian turboprop engines. There was a factory making stationary industrial gas turbines in Shanghai and probably several others we did not know about. Until recently, they had concentrated on military work and had all been part of a secret industry off-limits to foreigners.

The Chinese aircraft engine industry had begun with the importation of drawings, parts, machinery, and even foundry sand from the United States in 1940 to build Wright Cyclone Aeronautical engines in Chungking for the Chinese Nationalist air force in the war against Japan. MiG 15 jet fighters imported from the USSR were used by the Communist Chinese air force for the first time during the Korean War. In the 1950s, the Soviets brought jet engine technology to China. The factories we saw had been set up at that time to build engines of Soviet design. Chinese engineers and managers were sent to study at universities and technical institutes in the former USSR; most of the senior people in the Chinese aerospace industry speak fluent Russian. When Mao Zedong broke off relations with Krushchev in the middle 1960s, the Soviet technicians in China were ordered home. They packed up their drawings and technical data and left, leaving the Chinese to fend for themselves. In most cases, that simply meant continuing to do what they had been doing, trying to compensate with quantity of output for the absence of any new products.

Central planning of an economy is a great way to concentrate national resources. You can win big if you have selected the right program and execute it with skill; if the plan is a dud, you can lose equally big. Several times, the Chinese invested great sums of money and much effort in programs that had little chance of succeeding and in fact never did. One of the biggest was the program to build Spey engines in Xi'an under license from Rolls–Royce.

The Spey was introduced to China when the CAAC bought 13 used Tridents from Pakistan International Airways (PIA). The Trident was too small to be economical and competitive with airliners such as the Boeing 727 or Douglas DC-9 but it was more modern than any other airliner in the CAAC fleet, as was its Spey engine. The CAAC put the Tridents in service on internal trunk routes, including Hong Kong–Beijing and Guangzhou–Beijing.

In the mid-1970s, the Third Ministry needed a modern turbofan engine to be used in a new bomber under development for the PLA. One of its engine factories, the Liming (pronounced "Lee Ming") Machinery Company, was working on the development of such an engine broadly derived from Soviet technology, but it was clear that it would take years to bring a new design up to the levels of performance and reliability needed for a new aircraft. The Ministry decided that the new bomber would use the Spey. There was some technical logic to this. The engine was already in use in a British supersonic fighter, a British subsonic attack aircraft, and in the Tridents of CAAC. In fact, there were few alternatives. The Chinese engine was years from maturity: the Soviets had severed relations with China and could not be counted on to make technology available for a new engine; relations between China and the United States were still at arm's length, and the United States was notoriously jealous of its engine technology. The British, on the other hand, were quite ready to work with China (for similar reasons, Japan picked Britain at this time as its partner in developing a new commercial engine, the RJ500). An agreement was made to build the Spey engine under Rolls–Royce license. The Ministry, which owned and controlled the Xi'an factory as well as all the others, ordered two complete engines from Rolls–Royce, and 50 more in the form of parts kits to be assembled in China. Rolls–Royce sent drawings for making the parts at Xi'an and as many as 100 British technicians to teach the Chinese how to do it. The Chinese ordered machine tools and tools and dies from Britain to replicate the entire Rolls–Royce manufacturing process. I have heard that the Ministry spent $350–400 million to set up the program.

The plan contained a fatal flaw, one inherent in a command economy: no one had given much thought to the preferences of the customer. The Ministry controlled the engine factories and the airframe factory that would build the new bomber, but it had failed to get the PLA air force to commit to using the new aircraft with the

Spey engine. The air force simply balked. I do not know why the PLA air force took this position, but there was no doubt that it doomed the program. Although Xi'an tested the two engines from Rolls–Royce to calibrate its test cells and assembled at least some of the parts kits, as far as the world knows these engines never made it into active aircraft. Nor did CAAC use the Xi'an factory to make spare parts for the engines of its Tridents. All the elaborate equipment imported at great expense from Britain stood more or less idle. This was an important object lesson: the Spey precedent gave us some reason to fear that working with the Third Ministry would not necessarily convince CAAC to buy GE engines and might even antagonize the aircraft operators who would be our ultimate customers. We tried to delve into the background of the Spey incident with the people from the Ministry, but they were embarrassed and reluctant to discuss it.

The CAAC's selection process for new aircraft and engines also posed a few puzzles. When Henry Kissinger came to China in 1972, and President Nixon in 1973, they arrived in various versions of the Boeing 707, including "Air Force One." Air France, Lufthansa, PIA, and other foreign airlines began to fly into Beijing, and most of them used Boeing 707 airliners. The CAAC, which reported to the State Council or cabinet directly—not through the Third Ministry—decided to buy the Boeing 707. The reasoning was entirely logical, to avoid risk by following the example of leading foreign airlines. But again the logic was flawed. Even while CAAC was copying what they had ordered years before, the foreign carriers were getting ready to replace their 707s with the larger and more productive 747. CAAC negotiated for more than a year with Boeing and Pratt & Whitney and finally ordered ten 707s, receiving the last ten to move down the Boeing production line. The CAAC also bought one spare engine from Pratt & Whitney for each engine installed in the ten aircraft. Other airlines typically would have bought one spare for each four installed engines. The CAAC insisted that Boeing and Pratt & Whitney sell it a 100% reserve of spare aircraft and engines. The Americans did not mind complying with their customer's wishes.

At the end of the decade, six of the ten 707s were parked unused on the ramp at Beijing airport and the spare engines were in storage. That seems wasteful to us who have grown up in a capitalist economy, where idle assets represent an economic loss, investment must be depreciated, and there is an imputed interest cost to inventory. But apparently the Chinese were following a different economic logic. Perhaps the magnitude of the Chinese purchase was conditioned by the memory of how the Soviets had cut off support in the 1960s, by the notorious difficulty of getting spare parts from Soviet factories even in good times, and perhaps by the relatively short intervals between overhauls for Soviet engines. A fully used airliner flies about 3000 hours/year Soviet engines were at that time sent to the factory for overhaul every 1000 hours. If an overhaul takes a year or more, as overhauls in the former Soviet Union frequently did, it is quite sensible for an airline to have three or four spare engines for each engine installed. In contrast, the Pratt & Whitney engines in the 707 were sent to overhaul at intervals of 6000 hours or more, although frequently it was only selected parts of the engine that had to be overhauled.

The CAAC ordered Boeing 747s as soon as it noted how they were displacing the 707 in other airlines flying into China. By the time CAAC's 747s were delivered, the airline had learned enough from its own operating experience with Boeing aircraft and Pratt & Whitney engines and from the example of foreign airlines to order supplies of spare parts in the same proportion as the rest of the airline industry, and it was clear to the Chinese that more than half of CAAC's investment in the 707s was superfluous. Both the ministry responsible for building aircraft and engines and the airline responsible for their operation had made embarrassing capital investment decisions. To make the best of a bad situation, the Third Ministry was given some of this surplus inventory and encouraged to learn something useful from all the money that had been spent.

Two programs were started. The Ministry's aircraft factory in Shanghai designed an airliner called the Y-10, which used four Pratt & Whitney JT3D engines taken from the pool of excess spare engines at CAAC and looked very much like a Boeing 707 shortened by one-third. Seen by itself, it could have been mistaken for a Boeing. Three aircraft were built, one of them for destructive ground load tests; another crashed during a test flight, apparently because of lateral instability caused either by the aircraft's aerodynamic configuration or by an instability in the flight control system. Nothing more came of the Y-10 program.

At the same time, the Shanghai gas turbine factory built one or two replicas of the Pratt & Whitney JT3D engines. Again these were not exact copies, in this case being perhaps 10% bigger than the original. The Chinese tested them on the ground and boasted that their engines had more thrust than the original engines. They were quite proud of their feat and showed the engines to Pratt & Whitney and GE as well as to representatives of the American embassy. They probably intended to demonstrate that China was capable of building modern aircraft engines without the need for detailed drawings or manufacturing data from the United States. Western companies take a less benign view of having customers "reverse-engineer" their products. The U.S. government was not pleased. In any case, while the Third Ministry's laboratories and factories worked on these experimental programs, CAAC appeared to pay no attention and continued to order Boeing airliners with Pratt & Whitney engines. Some time later, CAAC began to order airliners with GE engines also, but I am getting ahead of my story.

Shenyang

Soon after our two technical seminars, we were invited to visit an engine factory with which the Third Ministry wanted us to establish a working relationship. In June, we were to go to Shenyang to visit the Liming Machinery Company. Shenyang is about 600 miles northeast of Beijing and north of North Korea. It is the capital of Heilongjiang Province, a part of Manchuria. Until the Russo-Japanese War of 1904–1905, Manchuria was part of the Russian empire and Shenyang was known as Mukden. Manchuria was Chinese until the Japanese set it up as a puppet state, Manchukuo, in 1932. Control reverted to China, after the Japanese defeat in 1945. Manchuria has coal and iron mines, many other minerals, and timber. It became one of the centers of heavy industry of China and of war production for the Imperial Japanese Army between 1932 and 1945. Liming, we were told, was one of the most important Chinese aircraft engine factories.

We took the overnight train from Beijing. A delegation from Liming Machinery Company met us at the station in Shenyang: the deputy chairman, the deputy chief engineer from the design bureau, and several manufacturing engineers. The deputy chairman, Cheng Huamin, was a man about 60 years old, an engineer who was also a deputy to the People's Congress. In background, experience, and his personal manner, he was an exact counterpart of the head of an engine company in the United States He would have felt quite comfortable running one of our factories. Cheng spoke a little English. Although he ran the Liming factory, he was only the deputy chairman; the chairman was a tough heavy-set politico, younger than Cheng and not involved in day-to-day factory operations. We would meet him at several formal meetings and banquets and several evenings of entertainment. He said very little and that always through an interpreter.

The deputy chief engineer had been educated at a Soviet engineering institute. The biggest part of his job was to give production engineering support to the factory, which was building 800 W5 turbofan engines per year and had the capacity to build as many as 100 engines a month. Liming also overhauled several hundred of these engines every year. The W5 was an old Russian design used in the Chinese A-5 attack aircraft, the Chinese equivalent of the MiG 19 fighter. The Chinese told us that the A-5 was of course a much better performer than the Soviet MiG 19. The engineering department also did some development work on improved components such as combustion chambers for the W5 engine and was trying to developing a new turbofan engine, the WJ5A.

We took a quick tour of the engineering building. Many of the engineers were in their 50s and 60s. Some of the older ones spoke a little English. The younger ones had been educated at the Northwest Polytechnic Institute in Xi'an or in Russian universities. The equipment in the test cells was relatively primitive. For example, the Chinese still used many banks of mercury manometers for measuring gas pressures in an engine, each manometer in the control room connected to a port on the engine in the test cell with a length of plastic tubing. Such instrumentation is fine for steady-state pressures but is neither precise enough nor quick enough in response to yield accurate results of dynamically varying pressures. We saw no sign of a major development program, nor the numbers of engineers and technicians one would need. Perhaps we were simply not shown such activities, but there was no sign of them anywhere, and the fixtures and instruments in the laboratories were very dusty. My diagnosis was that there was a lack of funding for development.

Employing more than 10,000 workers, Liming Machinery had the capacity to build more than 1000 engines per year and overhaul 800. In the United States, the major engine companies make only about 40% of the value of a complete engine themselves and contribute another 3% of the value in final assembly and test. The rest of the hardware, by value the bigger part, comes from a big network of specialized suppliers, some of which specialize in making certain parts, such as turbine blades, for all competing engine manufacturers. Life was different in China. The Liming factory made most of the engine parts needed for assembly of complete engines. Large forgings and some components such as ball bearings came from other factories owned by the Third Ministry, but Liming made almost everything else: small forgings and castings; many of its own machine tools; all of its dies, jigs, and fixtures; and even nuts, bolts, and small fittings. Such a degree of vertical integration, of self-sufficiency, was typical of Chinese aerospace factories, as it was of those in the former Soviet Union. The absence of market pricing gives suppliers in a socialist economy little inducement to meet their customers' needs because their first obligation is to satisfy the central planners. That gives each factory downstream the incentive to be as self-sufficient as possible so that it too can meet the production targets of a central plan. The process almost certainly creates higher than optimum costs.

We were a little taken aback by the pace of production in the factory, much higher than we had expected. By American standards, 800 engines per year means 640 for installation in new aircraft and 160 as spares. Six hundred and forty new engines for installation means an annual production of 320 new A-5 fighters. That seemed to us very high in comparison with the West; at that rate, Chinese military airfields would be covered with A-5 aircraft, a type whose performance was 20 years behind that of aircraft in service in South

Korea, for example, or in the Japanese Air Self Defense Force. Why was Liming building so many outdated engines? I can give only partial explanations.

First, one can build only what one knows how to build. The Soviets had give China drawings and data for the W5 engine but had walked away in the mid-1960s, leaving the Chinese with technology frozen in time like a fly in amber. Since then, the Chinese had continued to build the old design, while the USSR built and deployed much more modern aircraft on its European and Chinese borders. Liming had drawings for the Russian W5 engine, so that was what Liming continued to build, even though it was 20 years behind the times. There was no money and probably an inadequate base of technology to develop another engine. Later we heard about the development of a bigger engine, the W7, but even for that the pace of development was very slow.

Second, all orders and resources came from the Third Ministry in Beijing—production schedules, materials, budgets, operating plans, and money to carry them out. If the Ministry wanted Liming to build 800 W5 engines, that is what the factory did, regardless of whether there were aircraft for them, and probably regardless of whether the PLA air force wanted them. That is how *command* economies work. In a *demand* economy, the customer gives the manufacturer a sign of what it wants by placing a purchase order. In theory, the Third Ministry and the PLA air force held such discussions, but the fiasco of the Rolls–Royce Spey showed that what the Ministry had its factories build was not necessarily what the customer wanted. There was very little direct contact between Liming Factory and the PLA air force.

Finally—and we did not understand this until much later—Chinese engines, like the Soviet engines from which they were derived, were operated with very short intervals between overhauls, perhaps 150–300 hours. Their parts were designed for longer service lives but the designers and operators kept times between overhaul inspections short to ensure high reliability and safety of flight. In the U.S. Air Force, a fighter is flown for between 200 and 250 hours per year, a pace at which pilots maintain a high level of proficiency. If the PLA followed a similar regime, its engines would be pulled out of fighter aircraft and sent to the factory for overhaul roughly once a year.

Assume that it takes at least one year to get such an engine to the factory and have it overhauled and then returned to an airbase somewhere in China. At the factory, the engine is torn down into its detail parts. These are cleaned and inspected. Some will be usable, other will have to be replaced with new ones that have their own ordering and manufacturing cycle. One year is not very long for such complicated logistics. That means that for every engine installed in an aircraft ready for flight there was another at Liming for overhaul or on the way there or back. And certainly the airbase would want to maintain a tactical reserve of spare engines, at the American level one spare for every four installed engines, so that aircraft could be ready for operation despite unscheduled problems and maintenance. If this logic is close to what the Chinese actually used, an annual production of 800 engines would be split into 360 for installation in 180 new fighter aircraft (a more reasonable number than 320 new aircraft); 80 more engines as a ready reserve at airbases to compensate for unscheduled operating problems and engine removals; and another 360 engines in the overhaul cycle or in transit between the air bases and the factory. As long as there is no interest expense charged against the capital tied up in the inventory of spare engines, this is an entirely rational system of product support. In Marxist economics, interest expense is not an important factor.

This legacy of Soviet design practice, of short intervals between overhaul inspections, would also explain why CAAC had ordered one spare engine for each engine installed in the ten 707 airliners it bought from Boeing. In a Western market economy, large inventories of spare engines mean large investments and high expenses, so there is a strong economic incentive to minimize inventories, and engines are designed to stay on the aircraft for many thousands of hours. Components are designed so that they can be removed and replaced easily when they themselves give a built-in diagnostic signal that replacement is necessary. The rest of the engine stays on the wings. It is not unusual for an American commercial engine to remain in service for 10,000 hours, more than three years of typical operation. Millions of flight-hours of safe and reliable operation have validated the long inspection intervals themselves and the design practice on which they are based. As CAAC gained experience and confidence with its Boeings, it adopted the same longer inspection intervals used by other airlines. The CAAC loved such long-lived engines and no doubt so would the PLA; but long product lives and high reliability represent a threat to the factories. Operators need fewer such engines and there would be many fewer overhauls. A factory like Liming, designed to build 1200 W5 engines per year and overhaul several hundred more might need only one-third of that capacity with a more modern product. That was the situation we found when we arrived in Shenyang.

Making Parts in China

Our objective was to create a relationship that would induce CAAC to buy engines for its airlines from GE. We decided that we would place orders for parts with Liming, building up to a volume of about $1 million/year over three years. Although the Chinese wanted to start with much larger numbers, we were determined to proceed with care. We selected parts that fit with what we saw Liming make for its own production. We would have to provide drawings and specifications for making such components, and therefore they had to be for our civil engines or for marine and industrial engines. The U.S. government controls the export of aircraft engine technical data, with military engines under the jurisdiction of the State Department and the U.S. DOD. In 1979, neither was prepared to approve the release to the PRC of data relating to military engines. However, the Commerce Department controls the export of data for engines other than military, and we believed that it would approve release to the Chinese factory of data for nonmilitary engines. We also decided to furnish the raw material from the United States to eliminate that variable in the quality of the parts. Liming would machine the parts and inspect them, with the assistance of a resident GE quality engineer who could also spot-check samples of the production himself. To the airline, the ultimate user, GE would remain the guarantor of the quality of the part.

In 1979, Liming looked like an American engine factory of the early 1950s. The machine tools were old, although certainly capable of attaining the levels of precision required by our parts—not much different from the levels required for the Chinese engines. We wanted a good understanding of Liming's capability in terms of product quality, delivery schedules, and costs. For that we had to have an unambiguous common technical language. Liming wanted to learn from our technology and to earn hard currency from exports. The subcontracts would allow us to attain both sets of objectives.

Engineering standards posed an interesting problem. For example, turbine disks are inspected for surface and subsurface flaws left by the forging process with nondestructive processes, fluorescent particle inspection, and magnetic particle inspection. These are in universal use in Russia, in China, and the West. U.S. airworthiness certification authorities insist that the machines used in the process, and also the fluids and the chemicals, conform to a U.S. standard specification and that the inspectors themselves be tested periodically to certify that they can detect flaws of a standard size. Liming could have used their existing Soviet machines and processes, but we would have had to go through the long and difficult process of validating them to the U.S. standard. We thought that it would be simpler to install American machines with American chemicals and then certify that the setup and the operators met the U.S. quality standard. That turned out to be true as far as getting U.S. certification went, but we had not counted on the practical difficulties in China. Liming had first to secure permission from the Third Ministry to order the equipment from American suppliers and get an allocation of foreign currency. We helped Liming place the order with our suppliers. Then we had to wrestle the equipment through Chinese customs. It took more than one year before the machines were in place in Shenyang and the Chinese inspectors certified by GE. All this time, the Third Ministry and the managers of Liming worried that they ought to be cutting chips and making parts, only to be held up by what they saw as GE's nit-picking attitude to quality control. We asked Liming to set aside and segregate a small area in the factory for making our parts. To create the atmosphere of a showplace for production of very special parts, we had them put a fence around it, paint the floor white, and install much brighter lights than in other parts of the factory. We sent wooden boxes to hold parts in process in individual compartments, rather than having them piled on top of each other on the floor or a simple pallet. That not only reduced the incidence of nicks and scratches but, just as important, it also reinforced the attitude of precision and high quality. Finally we brought in white cotton shop coats embroidered with the "Monogram," the famous GE trademark, so that each person working on our parts could be identified as a member of the elite team.

The Chinese customs inspectors exploded. They had reluctantly accepted pressure from the Third Ministry to allow Liming to import machinery and tools for the work on its export orders for GE, but the shop coats were the last straw. They were not an essential element of the technical production process, Chinese customs decided, and charged us exorbitant import duties to bring them in.

Liming responded with good-natured enthusiasm to our coaxing and coaching. As time went on, they made a broader range of parts for GE and later also for SNECMA, GE's partner in CFMI. The quality met our specifications. Delivery was no worse and no better than that of any other GE supplier or of GE's own factories. And cost? In America we have our own accounting orthodoxies, and other people's ideas seem quaint to us. This was to be another lesson in the difference between Chinese economics and ours.

For each purchase order, GE negotiated a price for the value added by Liming. In the beginning and perhaps later as well, Liming had little idea of what its costs were. This was a consequence of socialist economics, of dialectic materialism. In a command economy, the central planners set targets for output, and furnish input resources of people, raw material, and working capital. Cost considerations play a minor role; notions of fixed and variable costs are not important (for American national programs such as the Manhattan Project, the development of the Minuteman ICBM, or of the Polaris submarine and its missiles, the U.S. management process was not much different from that in a centrally planned socialist economy). Yet even if they had no notion of what their true costs were, the Chinese understood price well enough. At first they asked GE simply to tell them what it would pay for Liming's work and would have accepted any number GE named. GE demurred. It wanted to be able to demonstrate in good conscience that it does not play favorites between one supplier and another, as well as to be free to use competition between its suppliers to exert discipline over their prices. But GE did teach the Chinese how to calculate costs.

On some parts, Chinese prices to GE were low. On others, typically parts with complicated manufacturing processes, they were high. In general, the Chinese prices were competitive. If GE warned Liming that it would lose the next order to another supplier with a lower price, the Chinese, just like suppliers in the United States, would lower their prices to retain the order.

There were very few cases where Chinese prices were significantly lower than those in the United States or Europe, even though the average cash wage at Liming in 1979 was only U.S. $23/month, less than $0.15/h. All Liming employees received free or subsidized housing from the factory, free or subsidized food, free medical care for themselves and their families, and free schooling for their children. Their taxes were negligible. But even if these benefits were worth ten times their cash wages, the total wage cost to Liming would still be only $1.65/h, compared with the cost of wages and benefits of at least $30/h at that time for a skilled machinist in the United States. Despite the dramatically low hourly wages, GE found few bargains in China.

There are several plausible explanations for this. The labor content of aircraft engine parts is a fairly low portion of the total cost; the direct difference in hourly wage rates is swamped by the cost of material, tooling and machinery, quality control, the meticulous recordkeeping required, and the cost of production-chasing across the Pacific and of an American quality engineer stationed in China. Furthermore, American parts manufacturers and their workers are very productive and competitive in cost. Some of the complicated parts that caused aches and pains when made on Liming's machinery could have been made quickly and effectively in the United States on modern machine tools controlled by computers.

Also, the Chinese quickly learned that price is set by competition in the market. If they believed that GE could get a part in the United States for X dollars, the Chinese would consider how badly they wanted the order and then ask a price a little lower than X. But not the 30% lower GE would have considered a real incentive for manufacturing in China. Some of those prices were available out of spare parts catalogs; others were common knowledge in the trade. Knowing that GE would always strive to maintain more than one source for each part, the Chinese saw no virtue in cutting their price significantly below what was required to keep the order.

In the end, each side accomplished a realistic objective but not its secret fantasies. The Chinese became trusted suppliers to GE and later to SNECMA, but they did not take over the entire market nor earn millions from exports. GE and SNECMA got a good supplier in China and built a good relationship with the Third Ministry but did not gain a great cost advantage from Chinese manufacture nor an overwhelming advantage in the Chinese market.

This last point bears emphasis. GE would have liked big cost reductions from parts production in China but even more hoped to gain preferential access to the Chinese market as a reward for its willingness to place some of the additional work in Chinese factories. However, the CAAC had different objectives. Like the Third Ministry, the CAAC reported directly to the State Council, the Chinese cabinet. The CAAC was concerned with operating reliable and economical aircraft and engines, which it wanted to buy at the lowest possible initial price, giving little weight to operating costs over the life of the product. It was notorious for the protracted "price auctions" whenever it held a competition for the purchase of new airliners. The experts in CAAC who conducted evaluations of competing aircraft paid no attention to production in China, or orders to Chinese factories from foreign bidders. They cared only about their immediate concerns. The Spey case made clear how little influence the Third Ministry had with the Chinese air force, and the Ministry was quite embarrassed about it.

In theory, the State Council would reconcile the conflicting Chinese objectives—that of CAAC for competitive aircraft and engines at the lowest price straight from the foreign factory and that of the Third Ministry for work for its factories and the resulting export earnings. In practice, the State Council allowed CAAC to order whatever it wanted from Boeing, which therefore had no incentive to become involved in subcontracts for Chinese factories, and then encouraged the Third Ministry to make whatever deal it could with foreign manufacturers.

Douglas, frozen out of the Chinese market, prevailed on the Chinese leader Deng Xiaoping during his first visit to the United States to approve a joint venture in Shanghai for the assembly of 26 MD-80 airliners for CAAC and the manufacture of a few parts. Douglas set up a production line in Shanghai, with as many as 90 Douglas technicians. Over a period of more than 10 years, the Shanghai factory assembled these airliners and delivered them to CAAC, which had been directed to accept them. When the production of the first batch was completed, another small batch was added.

The cost of building airliners at a rate of between two and five *per year* must have been tremendous. The Shanghai joint venture factory and Douglas almost certainly had to absorb the extra cost because we must assume that CAAC would refuse to pay more for an MD-80 assembled in Shanghai than for one built in Long Beach, California, particularly when CAAC had not wanted the aircraft in the first place. The CAAC put them in service on its internal routes. It is unlikely that the Shanghai program was profitable, but it served to create a presence for Douglas in China and keep Boeing on the *qui vive*. Airliners are big-ticket items, and in China as in many countries, their purchase requires approval by the government. The State Council began to press foreign suppliers to balance what they hoped to sell in China with what they bought there, even though the General Agreement on Tariffs & Trade (GATT) contains a Civil Aircraft Annex that prohibits a purchasing government's insistence on such "offset" purchases as a condition for the airline order. But China is not yet a signatory of the GATT, nor a member of its successor World Trade Organization. The presence of Douglas and the ambitions of Airbus Industrie to enter the Chinese market led Boeing to reconsider and place orders on Chinese factories for parts for its production in Seattle, Washington. But the bottom line remained: what CAAC bought was not necessarily determined by who bought what from Chinese factories. If there were concessions to be made during a competition, the most effective place to make them was to CAAC, not to the Third Ministry.

As the first American supplier to enter the Chinese market, Boeing could afford to play coy with the Chinese factories while concentrating on CAAC; Douglas and all those such as GE who had entered the Chinese market later had to use whatever links they could forge to get a foothold. Even if they suspected that it was not necessarily the deciding factor in an equipment selection by CAAC, they pursued the Third Ministry and its factories as potential allies. The Ministry did what it could to support the competitive proposals of the foreign companies that gave it work. These in turn worked hard to persuade CAAC that competition among suppliers would be good for the airline. As CAAC awarded orders to these foreign suppliers, the strategies of Boeing and Pratt & Whitney necessarily began to converge with those of the latecomers.

The Liming Visit to America

We had invited Liming's Vice Chairman Cheng Huamin to bring a group of his engineers and manufacturing people to our factories in Lynn and Cincinnati. They accepted, and we issued formal invitations through the Chinese embassy in Washington, D.C., as well as asking the U.S. embassy in Beijing to help with visas. When we were in China, the only time we spent money of our own had been shopping for souvenirs and for the banquet we had given at the guest house in Shenyang. All other expenses in China had been taken care of by the Third Ministry. It was clear that we would have to do the same for them during their visit to the United States.

They arrived in November 1979 and went immediately to the Chinese embassy, a huge building on upper Connecticut Avenue near the Washington, D.C., Hilton. There they stayed for three days. We then picked them up with our airplane early in the morning and brought them to Cincinnati. On the way in from the airport, we stopped for lunch at a small Cantonese restaurant, where we were joined some 30 minutes later by an official from the Chinese Consulate-General in Chicago. The man was tall, rawboned, and spoke perfect English. He had come by himself and was clearly the political minder for the group from Liming. He took a chair on my left. Vice Chairman Cheng sat on my right. The new man introduced himself.

"My name is Wang Defu. You should call me Mr. Wonderful," he said, and helped himself to some food. He asked me whether we had any concern about the stability of the Chinese government, in view of the repression of the Democracy Wall movement and the advanced age of many members of the Central Committee.

Until then we had avoided any mention of politics. But I saw no reason not to respond in kind. "The people at the top are very old, and soon will be dead. We wonder what will happen next." "No need to worry," Wang replied. "Our leaders are healthy, and most of them will lead a long life. In any case, they have arranged for an orderly succession. I have to tell you," he continued, "we have some concerns about the stability of the American government." Considering how the Iran hostage crisis had more or less paralyzed the Carter administration during the summer and mindful of the election of U.S. President Ronald Reagan a couple of weeks before, this was not as impudent as it sounded. "What brings our two countries together after so many years of hostility?" he asked. The other Chinese sat silent and transfixed.

I told him that there was a classic romance about China in America, going back to the clipper ship trade of the nineteenth century and to the American missionaries. There was also a Chinese romance with America, particularly in higher education—despite the history of the Chinese Exclusion Acts in California and the abuse of coolies brought here to build the transcontinental railroad. Each country saw the other as a large potential market, although both were likely to be disappointed (now, some 16 years later, the United States has an annual trade deficit with China of some $30 billion; at least one side has not been disappointed). Finally, we shared a common enemy, the Soviet Union. He certainly agreed with that. The others began an animated conversation among themselves, not translated into English.

Cheng Huamin spoke a little English. The interpreter who had come with him later went on to become the interpreter for Premier Li Peng. Some of Cheng's team, it turned out, had studied at American graduate schools. They understood English, and after a little practice could speak a few words. They were rusty after years of not having used the language, perhaps even having to conceal their knowledge of it during the Cultural Revolution. Cheng himself, who was a delegate to the People's Congress from Shenyang, analogous to a member of the U.S. House of Representatives, had been denounced and imprisoned during the Cultural Revolution. He was a tough and dignified gentleman, an experienced technocrat and leader. People of that caliber were likely targets for the Red Brigades, the teenage thugs sent out by Chairman Mao to destroy all symbols of authority other than the party and the army.

Although the group was halting in its use of English, most of them spoke Russian well. The younger men had studied at Russian universities, and of course Liming was producing engines of Russian design with data from the Soviet Union. All of that had come to a halt when Krushchev and Mao parted company in 1966 and the Soviets pulled out all of their technical assistance teams and documentation.

The Liming team stayed for three weeks, visiting the Lynn and Cincinnati plants and reviewing in great detail our manufacturing processes for the parts we planned to have them make. They felt quite at home in our factories, not so different from theirs, although we had more modern machinery and cleaner, better-lit workshops. We put much more emphasis than they on material handling and inventory control, not surprising because their inventory was free to them and we were charged interest by GE for our working capital. We bought much more of the content of our engines from suppliers, as much as half, where Liming made as much as 90% in its own factory. And we used fewer people. The work discussions did not present a major problem. GE had highly qualified engineers and scientists of Chinese ancestry, many of whom volunteered to help us as interpreters and explain details of the manufacturing process. Each side was very curious about the other.

Food was a problem. I had decided that we would serve our guests American as well as Chinese food, just as they had done for us in China. It was a good thing that we served at least some Chinese food. During a meal at a Chinese restaurant in Cincinnati, one of our people asked the visitors whether the food was authentic. "No," came the reply, "but it is a lot easier to digest than the strange American food." After three weeks on the road, their stomachs were letting them know that they were uncomfortable. Those of us at GE who traveled around the world can attest to the fact that Chinese food can be the salvation of the traveler.

The Missing Engine

Some time in the early 1980s, GE received an astounding message. The Third Ministry wanted to buy two CFM56-2 engines, engines built by the joint venture with SNECMA and used at that time in reengined Douglas DC-8 airliners and the KC-135. The Ministry had established a letter of credit for some $6 million for that purpose.

There were many prudent reasons for refusing such an order. CFMI sold engines to airline operators, which the Third Ministry is not. Two engines is a small sale, but even two were likely to need substantial engineering support from CFMI. The reason given for the purchase seemed hard to believe—to experiment with replacing the Rolls-Royce Spey engines in the 13 DeHavilland Trident airliners CAAC had bought from PIA. This seemed extremely unlikely from an engineering standpoint: by American reasoning, the idea of changing engines in 13 airliners was not realistic, even though two CFM56 engines replacing three Speys would save fuel and make less noise. The Chinese would have to design an entirely new tail section for the aircraft where the engines are mounted. The aircraft would have to be certified as if it were a brand new design. And if the operating results were successful, as no one doubted they would be, the other 12 aircraft would have to be modified in the same manner. All of this development, certification, and aircraft modification would cost a mountain of money, impossible to recoup from the savings in fuel, and perhaps not much less in total than buying 12 new 737s from Boeing.

So the reason given by the Third Ministry looked flimsy to us. The U.S. DOD office that deals with export licenses to China speculated that the Chinese had a different objective: to take a CFM56 engine apart and "reverse-engineer" a Chinese copy. The CFM56 was far more advanced than any engine in China. The Chinese had already reverse-engineered a copy of the Pratt & Whitney JT3D engines CAAC had bought for its Boeing 707s and had shown it, a little bigger and more powerful than the original, to Pratt & Whitney, proud of the fact that they could pulloff such a stunt. Whether the Chinese ever made more than one copy is not clear. China has been traditionally free and easy about taking other people's intellectual property and making copies of foreign products. The phrase "Chinese copy" is used to describe the uncritical way in which every aspect of the products was often copied. Even in 1998, China's entry into the World Trade Organization and China's trade relations with the United States are held hostage to the Chinese disregard for foreign patents and property rights. To Western manufacturers, these are serious problems.

The U.S. DOD had other concerns also. The CFM56 engine uses the same high-pressure compressor, combustor, and high-pressure turbine as are used in several advanced U.S. military engines, among them the engines for the B-2 bomber and F-16 fighter. The United States DOD wanted to make sure that the performance advantage of U.S. combat aircraft was not compromised or the performance of Chinese combat enhanced by reverse-engineering of the CFM56-2 airliner engine. I argued to GE's management and DOD that the Chinese could get access to the detail parts of CFM56 engines in the overhaul shops of a number of airlines around the world. Sooner or later they would be able to buy airliners equipped with CFM56 engines themselves and peek inside. We were trying to protect a real secret, but a small one, and it would be compromised soon in many other ways. What the Chinese were doing, I believed, was to test GE, SNECMA, and the American and French governments to see whether we were collectively serious about doing business with them. After a lot of debate, and with the letter of credit about to expire, CFMI received U.S. and French government approvals to make the sale. There was a condition attached: the Chinese had to agree that GE and SNECMA could inspect the two engines on demand. The trading company of the Third Ministry, the China Aero-Technology Import & Export Corporation (CATIC), agreed, and took delivery.

In hindsight, I believe that we overlooked an important Third Ministry objective. When we judged Trident reengining as unrealistic, it was by American economic standards. Quite likely these did not apply to this situation at all. The Ministry was in charge of design and production of aircraft and engines. It was in the business of developing, designing, and building experimental aircraft (one such was the Y-10, a four-engine airliner that looked like a two-thirds scaled version of the Boeing 707; the Y-10 never went into production). Perhaps the mere experiment of reengining had value to the Third Ministry—no matter how much it cost and whether all of the Trident airliners were ever modified.

Nothing ever came of the Trident reengining program, although it was never declared officially dead. On one occasion our people in China requested permission to inspect the engines and were allowed to see them. After that, the engines disappeared. There were all kinds of rumors, including one that one of the engines had exploded during a ground test. Just in case the Chinese had disassembled one of the engines and could not get it back together, we offered obliquely to help if they needed it. There was no reaction. The U.S. DOD of course was furious. For years after that, the export control people held it against the Chinese and against GE that the Chinese had not lived up to their commitment in the export license as well they should have.

Several years later, CATIC bought an American aircraft parts manufacturer in Seattle, Washington, a supplier to Boeing, paying an exorbitant price for the company, far higher than its order book and profits would have justified. Such investments are subject to the approval of a U.S. government agency, the Committee on Foreign Investment in the United States (CFIUS). The U.S. Treasury Department is in charge of CFIUS, and most other cabinet departments are represented. The committee has set special conditions to safeguard U.S. interests in several foreign acquisitions of American companies but had never recommended against a foreign investment until the CATIC purchase of the company in Seattle, Washington. The U.S. government refused approval. CATIC was now forced to unload its acquisition under the worst of circumstances. Prospective buyers knew that CATIC would not be allowed to retain ownership, and the true worth of the company to an American buyer was a lot less than CATIC had paid. CATIC suffered a huge loss of face. The vice president of CATIC who had made the purchase told me angrily in Beijing that he planned to take his lawyer to see the chairman of CFIUS and would threaten to sue him for withholding approval without reason.

I tried in my work not to confront the Chinese, with whom we were trying to build a good working relationship. The Chinese particularly among Asian people do not react well to threats. But I saw no reason to listen without comment to threats from a man who had been humiliated by the consequences of his own action, and told him that CATIC would no doubt take whatever action it found appropriate but that in my opinion he would make a mistake by threatening the U.S. government, adding that CFIUS was not acting capriciously and that the U.S. government was sending CATIC a message: if China did not live up to its side of a bargain, as in the case of the two CFM56 engines, the United States would take steps to protect its interests. The conference room became very quiet. After a minute, the president of CATIC said quietly, "That is exactly what Mrs. Chennault told me in Washington." Anna Chennault is the Chinese-born widow of Lt. General Claire Chennault, the commander of the American Volunteer Group and after that of the 14th Air Force in China during World War II. She is well connected in China and Washington, D.C., and had been acting as a consultant to both sides.

There was one more odd twist to the case of the two engines. We had engaged Everbright Corporation, a trading company in Hong Kong owned by the Chinese government, as marketing consultants to persuade CAAC to buy two 737s with CFM56 engines, and we had urged them to add resources to their technical staff. Everbright in fact hired an experienced engineer from CATIC and sent him to Cincinnati to help translate the CFM56 brochures into Mandarin. When we were done, a GE chauffeur took him to the airport for his flight to California and back to China. Somewhere after we dropped him at the airport, he disappeared. Everbright announced that he had defected and taken a large sum of money with him. They asked GE to help find him. He must have gone underground somewhere in the United States as an illegal immigrant, surely not the first. We were not in the private detective or immigration law business and told Everbright that we knew nothing of him after our chauffeur left him at the airport. There was little more we could do to help them. But I have some sneaking sympathy for the defector.

Setting Up in Beijing

Although CAAC had yet to show much interest in our products, the Third Ministry was obviously anxious to have us do business with its factories. GE is often seen as a source of technology and expertise, and the Chinese could see the benefit to them of competition between Western engine firms. To reach CAAC, GE Aircraft Engines needed to set up an office in Beijing.

I picked one the many Chinese engineers at GE Aircraft Engines to set up our office. His name was Sung and he had emigrated from the south of China with his parents in 1949. Sung was overwhelmed and delighted with the opportunity. Overseas Chinese were often anxious to visit China and the Chinese certainly court them as the rich relatives from abroad, but always as members of the family. Sung was excited about his forthcoming assignment to China. While we waited for his passport, he went to San Diego, California, to deliver a technical paper. After the conference, he and a group of friends went to the beach for a picnic. Sung went for a swim and was never seen again. His clothes were left on the beach, neatly folded with his wallet in the pocket of his trousers.

There is a notorious undertow along the shore. The sheriff sent out searchers and a helicopter, but no trace of the body was ever found. Sung's family hired a psychic with some reputation in matters like this, but to no avail. Sung's wife, who spoke little English, waited in Cincinnati, Ohio, with two small children, receiving no benefits from her husband's GE life insurance because the body had not been found. Our personnel people managed to correct that quickly. At that point, Sung's passport arrived from Washington. It would have been a hot commodity on the black market. I turned it over to the GE lawyers.

Now what were we to do? Our personnel people recommended Walter Chang. When he returned to GE, he had no interest in going back to compressor design, and the personnel people had assigned him to the Airline Marketing Division as a sales support engineer, working with the GE salesmen who dealt with customers. All they expected from Chang was detailed nuts-and-bolts support, when what he wanted to do was to move into a position where he could exercise some initiative. The head of personnel and others at high levels, impressed by his enthusiasm and energy, wanted to find him an assignment in which he could use his potential and sent him to me as the man to open our office in Beijing. It was a decision that would lead to dissension inside GE, and between us and the Chinese; I describe it at length to illustrate some of the problems of setting up an isolated sales office in what had been an isolated country.

Chang and I traveled there together, accompanied by two engineers who would discuss the CFM56 engine with CATIC and CAAC. He was very excited and I was a little nervous when we arrived at the airport in Beijing for his first visit to China since he had fled with his parents in 1949. If the Chinese passport people gave him any problem, we had agreed among ourselves, we three "Europeans" (two Americans, one Frenchman) would make a scene until the Chinese either allowed Chang to come with us or deported all of us together.

The three of us were waved through with barely a glance. Chang was stopped for a lengthy discussion. I held our group there even as the Chinese motioned to us to go through to the luggage carrousel. We waited. Finally Chang joined us, nonchalant and cheery, and we went through. The immigration officer had noted that he had been born in Beijing, and they had chatted lightly about relatives in China and life in the United States, Chang said. Nothing to be concerned about, but thank you anyway for making such a show of solidarity. I am still convinced that the conversation was less casual on the side of the passport official than Chang may have believed or made it sound.

In Beijing, Chang began the milieu with zest. He visited the hospital where he was born. He argued politics with the sullen floor boys, male and female, in the Beijing Hotel, and he showed an irrepressible talent for taking over meetings with our Chinese customers. I had brought presents to hand out during major banquets, dignified little souvenirs from the gift shop of the Cincinnati Art Museum. Walter took over, giving his own presents of ball-point pens and foam cans of shaving cream with a heating device in the lid, certainly more amusing and probably more appreciated.

Chang set up his office in the Beijing Hotel, taking a suite in which he lived in the bedroom and used the large sitting room for meetings. Suite 1735 is now unoccupied, kept as a memorial to the late premier Chou Enlai, who had used it occasionally years before

GE appeared on the scene. He also rented two rooms on the floor below, one of them as a store room, the other as both an office and a home for his secretary. She was a Chinese citizen, married to a man in Hong Kong, very capable, arranging meetings and making reservations, and acting as Chang's deputy whenever he was away from the office. Two years later, she announced that unless GE gave her a big salary increase and an occasional business trip to the United States, she would look for another job. Chang explained to me that this was the negotiating style in Hong Kong and that there was a lot of competition from foreign companies looking for capable office managers fluent in English and accustomed to Chinese bureaucracy. The GE personnel people in Hong Kong were not very flexible in administrative matters, and I too was not pleased about what I saw as pressure. We did not accommodate her, and she went to work for Boeing, which also had its office in the Beijing Hotel.

There was then little choice of office location for a foreign company other than in a major hotel. In 1979, there was almost no other office space available anywhere and certainly almost no housing for foreigners. Many of the rooms in the Beijing Hotel were occupied permanently by representatives of Japanese trading companies. As more and more American and European companies came piling in, space grew short. Telephone and facsimile connections, difficult anywhere, were easier in the major hotels. The hotels realized that they were in a seller's market and began to abrogate leases and raise rents. We were fortunate that we got in relatively early and had that feisty native of Beijing, Walter Chang, to negotiate on our behalf.

Soon the other GE businesses in Hong Kong concluded that they had to have a presence in Beijing. By that time, they were able to rent office space in the International Club, and they suggested to Chang that he move the GE Aircraft Engines office from the Beijing Hotel to join them. He did not agree. His Chinese customers, he argued, felt at ease in the Beijing Hotel, free to come and go with other Chinese; they would feel much less comfortable visiting an obviously foreign location in the International Club. There were restaurants in the Beijing Hotel for easy entertaining. Boeing had its office there. There was no real advantage to his being in the corporate office with all of the other GE operations, which had nothing to do with our business. These points were valid, but the people in the corporate office interpreted Chang's refusal to mean that he, and perhaps GE Aircraft Engines to the extent that we supported him, did not care to be associated with them, did not care to be a part of the GE team. And perhaps there was some truth to that. Relations between the two offices deteriorated rapidly because neither had much use for diplomatic niceties. One of the other businesses, GE Plastics, also set up a separate office but that seemed to cause much less tension.

The corporate office was run by Richard Abington, a retired U.S. Navy pilot and an ambitious man. Jack Welch, the GE chairman and CEO, told Abington that he was holding him responsible for GE's success or failure in China, and that he should make sure that only those GE businesses with a real chance of success entered China, keeping out those that were merely playing in another exotic market. Abington ran the Beijing corporate office from Hong Kong, where he lived and kept his office, visiting Beijing once every six weeks. He appointed a deputy for Beijing, John Wu, the locomotive salesman for the GE Transportation Systems Division. GE sold several hundred diesel-electric locomotives to the Chinese railways, which were still using thousands of steam locomotives. There was fierce competition for the locomotive market from General Motors, Hitachi, Siemens, and Alsthom. Prices were cutthroat and profit margins vanishingly slim. Another Chinese–American, Mae Mitchell, a physicist from San Jose, California, was sent to Beijing to act as administration manager. GE salesmen came into Beijing from Singapore, Hong Kong, or the United States and worked out of the Beijing office for weeks at a time. Some of the secretaries were brought in from Hong Kong and lived in the Min Tsu Hotel. Other secretaries and chauffeurs were hired from the Chinese government office that was the sole authorized employment agency for foreign companies. GE paid the equivalent of Hong Kong wages to this agency, which sent over employees whom one could accept or reject. They were paid a much smaller salary by the agency, usually supplemented by special benefits from the foreign employer. Presumably they also kept an eye on their foreign employers on behalf of the Public Safety Bureau, the Chinese counterintelligence police.

Walter Chang was on bad terms with John Wu and Mae Mitchell almost immediately.[12] Wu had been told by Abington that he was the senior GE manager resident in Beijing, more or less in charge of all other GE people there. Walter was not prepared to play second fiddle to Wu in any way. No one from GE Aircraft Engines had directed him to report to John Wu, and Walter joked that he himself ought to be the senior GE man in Beijing because he had set up his office first. The rest of GE did not smile. Mae Mitchell took her cue from John Wu and made Walter's life difficult with administrative details about his leases for office space and cars, his hiring of a secretary, and his expense accounts.

Chang and I talked about the problem. I advised him to pay political respect to the corporate office in Beijing. The GE trademark was a powerful resource, and we needed to save our energies for working with customers rather than squabbling with others about who was the most senior. "Keep doing your work," I told him, "but behave like a civil and cooperative member of the GE family." He was disappointed. He was sure that he was acting in the best interests of GE Aircraft Engines and expected me to back him up against these unreasonable demands from the corporate people. Abington was also dissatisfied with my approach. If GE Aircraft Engines did not subordinate itself to his office, then we did not understand the mandate from the chairman. Abington and I had two meetings, one in Hong Kong and one in Cincinnati, Ohio, to try to resolve the argument. I acknowledged the role of the corporate office in Beijing as the symbolic face of GE, responsible for integrating the interests of the different GE businesses active in China. I also emphasized that Walter Chang and the representatives of the other businesses had the primary responsibility for marketing their products and that they were accountable for that to the management of their businesses in the United States. These were sensible ideas and tactful

---

[12]After touring China with Walter Chang, the Vice President heading GE's Military Engine Division criticized the location of Abington's office in Hong Kong, far from the customers in Beijing. Abington of course blamed Walter for the criticism.

words, but nothing came of them—Abington insisted that he was in charge, responsible to GE's chairman for what went on in China, and that Walter Chang was not a team player and ought to go.

In 1985, Brian Rowe, the president of GE Aircraft Engines, was to come to Beijing to meet with Aviation Industry Minister Mo and Prime Minister Li Peng. We were in the midst of preparations for this important event when we received disturbing news. Dick Abington had sent a six-page handwritten letter to the executive at GE corporate headquarters responsible for the GE International Sector, reporting a serious matter. A GE consultant in Beijing, an old and respected retired government official, had summoned John Wu to his house. The Bureau of Public Safety told him that a certain GE employee, of course Walter Chang, had engaged in activities hostile to the interests of China. If Chang did not mend his ways, he might be arrested and thrown in jail, and GE risked being expelled as a company. The Public Safety people asked the old man to pass that message on to his client GE. John Wu flew to Hong Kong at once to inform Abington, who consulted others in the American community, all of whom advised him to take the warning very seriously. Abington recommended that GE Aircraft Engines pull Chang out of China at once.

All of us agreed that the Chinese warning had to be taken seriously. We needed to deal also with Abington's attitude that Walter Chang was an "unguided missile" and not a team player: the Chinese warning appeared to vindicate his judgment. And while we dealt with this unexpected crisis we needed to proceed with the preparations for Brian Rowe's visit. After some discussion with corporate headquarters and Abington, we agreed that we should continue planning the visit. There was no immediate danger, we concluded, and Walter Chang could be allowed to continue the preparatory work. Dick Abington as GE's "national executive" for China would naturally participate in the visit. But we also had to take action on the Chinese warning, both because of its importance and because Abington would continue to insist that Chang's indefinite presence in China was bad for GE as well as a provocation to the Chinese. Even GE Aircraft Engines' chief technical representative in China, a retired U.S. Marine, agreed with that, Abington told us.

I was in Lynn at the time preparing for the visit and asked Chang to come to the United States. After we had gone over plans for Rowe's visit, I told him what had happened. He was thunderstruck. He refused to believe that there could have been such a message from the old consultant, whom he had never met. He refused to believe that there could have been such a message from the Bureau of Public Safety. He assured me that he had never done anything that could be construed as an action against the interests of China. He told me that he would confront the consultant and the police and challenge them to come up with anything derogatory. He was sure that this entire matter was an invention by a jealous John Wu, seized upon by Abington who had always resented Chang's independence from his direct command. He announced that he would sue Abington for character assassination. And so on and so forth. It was easy to understand why he was upset. Neither he nor Abington was prepared to concede for a moment that the other might be acting in good faith. I warned Chang that while he was free to do whatever he wanted to protect himself, I did not want him to harass the old consultant nor to sue Abington. Either action would be hostile to GE Aircraft Engines' interests in China. GE Aircraft Engines would support him as its key man in China, but not if he were to go against our interests. In the meantime, we had to take care of Brian Rowe's visit.

Rowe arrived in Beijing in his Canadair Challenger executive aircraft, accompanied by the airline marketing division vice president and their wives. The plane was directed to a different terminal from the one we expected, and an undignified scramble got the reception committee and cars and vans for luggage and for the crew all to the right place. Understandably, nerves were a little taut.

The cars and vans took us to the Diaoyutai, the state compound where important guests of the government were put up. Presidents Nixon and Reagan had both stayed there. But, in 1985, it had become possible for private parties to get accommodation there if they knew how to arrange it. Chang knew how. We were all in one huge villa in the walled compound, surrounded by a fine garden, in suites of more or less splendor according to rank. The telephone service was unusually good, although a call to me from Cincinnati, Ohio, in the middle of the night was first put through to Brian Rowe's suite because he was the most important of our group. "How were you able to find us?" I asked the man who called. "No problem," he replied. His wife, who is Chinese, placed the call and asked the American long distance operator for the GE party in the Diaoyutai. The Chinese operator knew exactly where we were.

Abington joined us from his hotel and, in a procession of long black limousines, we were driven to the office of Prime Minister Li Peng in the Zhongnanhai (the name means "Middle South Sea," referring to its location in Beijing and the ornamental lake in the compound), where the government and party bigwigs have their offices and residences. We were taken to a pavilion probably separate from Li's normal office. There on the front steps, with protocol men from the foreign liaison department of the Third Ministry, was Walter Chang. We were all ushered into a splendid conference room and greeted by Premier Li and his entourage, including Minister of Aviation Industry Mo. I recognized their interpreter as the same woman who had accompanied the Shenyang group to Cincinnati in 1979.

Brian Rowe was invited to sit on Li's right, between him and Minister Mo. The rest of us arranged ourselves to the sides and behind in two rows. Conversation would be limited to the principals. Tea was served. Li welcomed us and complimented Mr. Rowe on the fine work GE was doing with the Liming factory. He hoped that GE's cooperation would be expanded to other factories as well. Then he talked about China's investment priorities, particularly in power generation. Li is an electrical engineer who graduated from a Soviet technical institute as a specialist in electric power generation. Rowe had discussed with Abington what business opportunities he should emphasize, and responded with an account of GE's broad capabilities, in power generation products as well as many others of interest to China, such as locomotives, plastics, and medical x-ray machines. He then added that our most important objective was to have CAAC adopt GE engines for its airliners. Li agreed that the Chinese would give serious attention to that. Minister Mo agreed and said that GE's cooperation with Chinese engine factories would certainly help it happen. He also asked Rowe to help the Ministry's engineers improve the fuel consumption of the WJ5 turboprop engine, then still widely used by CAAC in its Y-7 airliners (the Y-7 is a derivative of the Soviet Antonov An 12), and Rowe agreed that GE would do so. Then Mo looked directly at Rowe and added, "Mr. Rowe, you and your company are certainly lucky to have such a good representative as Walter Chang here in Beijing. We find it very easy to work with him, and he is doing fine work to bring your company and us together." Li Peng stood up. A photographer came to take our picture with him. We made our farewells and left, satisfied with the positive tone of the meeting.

That evening we offered a banquet at the Beijing Hotel. Chang and his secretary had engaged the hotel's great banquet hall and put the catering staff to work. Several hundred Chinese government and industry people had been invited, as well as members of the U.S. embassy staff and others in the foreign community. As they arrived, guests were invited to sign their names—in Chinese characters if they could—on a long fabric scroll, using a calligraphy brush and black ink. Round tables for 10 had been set up inside, more than 30 of them. There was one GE person at each table, wherever possible, also with someone able to interpret between Mandarin and English. At one side of the hall stood a lectern with a microphone and floodlights for the photographers and television cameras. A great red banner hung high on the wall, proclaiming in gold letters in English and Chinese, "GE Aircraft Engines welcomes its friends and partners from the Chinese aerospace industry!"

We found our tables and went through the motions of introducing ourselves, whether we could understand each other or not. Waiters poured orange pop and beer and plum wine and mao tai and began to bring platters of food to be arranged on the lazy susan in the middle of each table. Hosts served the guests to their right and to their left and others began to help themselves. Conversation, no matter how halting, began to buzz through the hall.

When the time came for the ritual toasts and speeches, I went to the microphone, Chang at my side to interpret. Floodlights came on, with the result that I could see nothing in front of me other than the microphone. More than a little nervous, I welcomed our guests and introduced Brian Rowe at the table next to the podium. At his right sat Minister Mo, the guest of honor. Rowe came to the microphone and introduced Abington as the senior GE executive in China, an introduction I had forgotten to make. He then spoke briefly and gracefully about GE's objectives in China, and about our hope that we could create a relationship of benefit to both parties. Minister Mo rose to make the response. As he had that morning in the premier's office, he applauded our initiatives with the Chinese factories. He mentioned the high reputation of GE's technology, and looked forward to the adoption of our aircraft engines and other products in China.

Then he continued, "It is particularly gratifying to us that so important an American enterprise is represented in China by Walter Chang, born in China and able to act as the perfect informed intermediary between GE and its Chinese customers." It was a stunning endorsement, almost as if Chang had orchestrated it. Perhaps he had, but even that could be read as an endorsement. There was obviously no cause for us to worry that he might be thrown into jail during Rowe's visit.

The minister proposed a toast to GE and left soon after that. In the custom of Chinese banquets, other people rose with their glasses in hand and went from table to table, toasting with friends and acquaintances. As the banquet broke up into clusters of individual conversations, Brian Rowe and his entourage met with Abington in a sitting room off the side of the banquet hall, to discuss the issue of Chang and his relationship with the customers and with GE China. Rowe challenged Abington with Minister Mo's praise. Abington agreed that this was remarkable but countered that it did not make two big problems go away: whether we liked it or not, the Public Safety Bureau had given GE a stern warning about Walter Chang, and Chang was not and never would be a team player with the rest of GE, unless Brian Rowe himself forced him to be. Rowe praised Chang's work in support of GE Aircraft Engines. Abington expressed some reservations about the results and did not back off his position that Chang did not intend to work as part of Abington's team, as Jack Welch had decreed every GE person in China should. Welch, of course, was Rowe's direct superior. Everyone held his temper, but Abington had not given any ground under considerable pressure.

In the car on the way to our next meeting, or sightseeing, or wherever we were going, Rowe said to me that although it was clear that Chang was very good as an intermediary with the Chinese government, he certainly was not able to get along with the people in GE China, and that I should redesign his job so that we could help him save face and make use of his evident strengths but get him out of permanent residence in China because his presence there caused so much friction.

I tried. We created a position called "Vice President—China Market Development" and assigned Chang to it. He could be based in Boston, where he could be with his wife and daughters for more than one week every couple of months. He would advise me on China market strategy and act as the contact with the Chinese government and as the scout for new business opportunities. He thought about it. He recognized that this was not a promotion but believed that he would retain some degree of freedom for marketing in China nevertheless. He also offered me his services in Taiwan and Thailand; the Thai air force chief of staff, Walter told me, had been a classmate of his. I turned him down on both. I believed that we should not use a single person to market both in Taiwan and on the mainland; each customer would distrust him to an equal extent. Besides, our marketing man stationed in Thailand would have fought tooth and nail against what he would have perceived as amateur help parachuted into his territory.

I replaced Walter Chang in Beijing with Robert Linton, a veteran of GE Aircraft Engines whose most recent assignment had been as GE national executive in Pakistan. Linton offered several advantages. He was tough and calm and able to work with Abington without giving up his independent link to us in GE Aircraft Engines. His assignment in Pakistan was an asset in that regard. Linton had also put in time with field service engineering and was able to establish an easy working relationship with that group. As an experienced field marketing man, he knew what the factory sales people wanted from the Beijing office, although even he had his troubles with them. Against that, he was not Chinese and did not speak Mandarin, which no doubt cost us some effectiveness for a while. The language gap was the most serious problem.

Chang, of course, was not mollified. Linton moved the office into the International Club with the rest of GE, which Chang saw as a repudiation and a mistake, continuing to argue that the customers preferred to go to the Beijing Hotel (where I had agreed that he could keep a room) rather than to an overtly foreign location. Abington also was not mollified. We could never rest safe as long as Chang was allowed to show up periodically in China and submerge doing heaven knows what. And besides, what was Chang supposed to do that Linton would not do also? Had Chang cooperated smoothly with our airline marketing people, we might have made it work. But there was a long history of clashes between them, aggravated when they hired an American–Chinese of their own who was as convinced of the rightness of his approach to CAAC as Walter Chang was convinced that it was wrong.

Conscientiously he looked for markets where he could make a contribution. Through his former Republic of China Air Force connections, he had learned that the twin-engined indigenous defense fighter (IDF) designed in Taiwan with help from General Dynamics was short on performance because the two modified Garrett Turbine Engine Company business jet engines lacked the required thrust. That had been a consequence of politics. In the Shanghai Communique of 1973, U.S. President Nixon had agreed that the United States would not provide new military equipment to Taiwan with greater capability than was already in place. Because the Garrett engine was a civilian model designed for executive jets, and the conversion to a fighter engine was being done by the Taiwanese in the Chung Shan Institute of Science & Technology, the United States was complying with the letter of the communique. Such genteel evasions are not unknown in diplomacy and are quite common in an Asian environment that generally shuns blunt confrontation.

Walter Chang looked over GE's range of products, picked a compressor and turbine used in the engine of the F-18, combined it with the fan of another model, and proposed a single engine that would replace the two Garrett engines and still give better performance in the IDF. The Taiwan Air Force took the proposal seriously and GE put a team together for the cooperative development and production of the engine between Lynn and Taiwan. Garrett was alerted by its allies in Taiwan, whose reputations were committed to making the IDF a success in its original form, and a full-scale competition was under way. The Lynn project team wanted no help from Chang, who had thought up the entire idea. To be fair, the people in Lynn wanted no help from anyone on my team, not only Chang. But we could not find a way of making use of his talents and energy. Then politics intervened once again: the improved version of the IDF was scuttled when, during the 1992 U.S. presidential election campaign, President Bush approved the sale 150 F-16 fighters to Taiwan, in an attempt to win votes in Texas and California where the F-16 was made. The diplomatic excuse was that mainland China had received Su 27 fighters from Russia. Once Taiwan had the assurance that it would get the F-16, it lost all interest in improving its own fighter. The Garrett engine was good enough and the improvement project was dropped. Soon after that, disappointed in GE and its irrational treatment of his best efforts, Chang took a leave of absence followed by early retirement. Another of my Asian employees, a Korean of great ability and even greater ambition, took note of that and wondered whether any Asian could make a career in a company like ours.

I saw Bob Olson, our original marketing man in China, somewhere along the way and with little satisfaction told him the story. What mystified me still, I told him, was the apparent discrepancy between the Public Safety Bureau's warning to Abington and the flowery encomium from Minister Mo. "That's not difficult to explain," Olson replied, and at last I learned what had happened (this account was later confirmed by Walter Chang; I have not asked the Bureau of Public Safety for its version). During one of the early negotiations between our Marine & Industrial Engine Division and the Chinese navy, the Chinese negotiators announced that they would have to recess for three weeks. "What's wrong?" our people asked. "Nothing," came the reply. Americans do not like to twiddle their thumbs for three weeks, particularly not in China: anxiety sets in quickly. "Where is the chief negotiator?" they asked. "He's off the scope for some time," they were told. Walter Chang snooped around and found out that the chief negotiator was hospitalized for eye surgery and incidentally that his birthday was approaching.

"Let me take care of this," Chang told the GE negotiating team. He located the VIP naval hospital where the official was confined and went there with a rose bush in a pot because he could not find a bouquet of flowers. Outside the hospital, he gave a hospital orderly a few dollars to trade clothes with him. Now dressed as a Chinese military hospital orderly, Chang entered the hospital with his rose bush and went from room to room until he found the negotiator, who was surprised to hear his voice but very glad to have a visitor. Chang explained that the plant was from the GE negotiating team, which was very concerned that the operation would abort the negotiation. "Tell them to be patient," the official said. "We'll buy the engines as soon as I get out." He made a telephone call, and the negotiations resumed the next day.

But they were not yet concluded. The Chinese side kept asking for more concessions, while the GE side declared that there was no more to be given. The negotiations were at a complete impasse, and the Chinese gave no sign that they were under any pressure of time to reach agreement. In fact, the admiral who was the ultimate decision maker had left Beijing to participate in the U.S. Navy's port visit to Shanghai, the first since 1949. Walter found out that after the American ships had left, the admiral had returned quietly to Beijing. Walter directed the GE negotiating team members to lie low in their hotel, as if they had left for the United States. Then he went to the military compound that contained the admiral's apartment. Chinese were admitted to the compound, he noticed, which was off-limits to foreigners. That evening he lingered outside, dressed in blue trousers and a white short-sleeved shirt, carrying a cheap plastic briefcase and with a lit cigarette dangling from the corner of his mouth, looking like any other of dozens of Chinese waiting at the gate. When the gate opened and the others entered, Chang went with them. He found the admiral's apartment and was admitted. The admiral, in pajamas, was taken aback. He told Chang that for his own protection he would have to report his presence to the Public Safety Bureau. They talked briefly about business: Walter told him that GE had made its best offer but that the negotiations were deadlocked. Only the admiral could move them ahead, and if he did the GE team would be on the scene within 24 hours. The admiral promised to intervene and Chang left.

To his dismay the main gate was closed. A small side door led past a manned sentry box. For a moment Chang considered climbing over the wall but rejected that quickly. The risk was too high. At that moment another visitor came hurrying out of one of the buildings and ran to the sentry box, where the sentry berated him for overstaying his pass. Chang sidled up and stood silently behind the man during the militiaman's tirade. Finally the sentry waved the man through and Chang silently walked out right behind him. A few days later, GE and the Chinese signed the contract for the engines. At the celebration banquet, the admiral, sitting in his uniform next to Walter, joked that only his wife and Walter Chang had ever seen him in his pajamas.

Clever, aggressive, creative marketing, you say? So it is. But if you were the Public Safety Bureau, you might take a very different view of an American, able to pass as a native Chinese, bribing an orderly to get access to restricted military facilities for the purpose of making who knows what arrangement with a senior naval officer responsible for a major purchase from a foreign company. Just think of it in terms of an American admiral, Bethesda Naval Hospital, and the Federal Bureau of Investigation.

Why did we not know any of this at the time? Why didn't Chang say anything about it then? And would we have done anything differently had we known?

The New Team

Because Bob Linton did not speak Mandarin when he arrived in Beijing (he made decent progress in understanding and speaking basic Mandarin in the four years he spent in Beijing; his next assignment, in Israel, was equally challenging linguistically), he needed some help, and we had to enlarge our staff in the Beijing office to handle the expanding workload. We hired two men in the United States and one in China. The first was Paul Sun, an engineer with GE's nuclear division in San Jose, California, which was scaling back staff because of the impossibility of getting United States licenses for the construction of nuclear powerplants. Paul's and his wife Alice's families had fled from China. Both of them had been born and educated in Taiwan, Paul later earning a Ph.D. from the University of Michigan. He was delighted to go to China; Alice worked as a certified public accountant and was a little less delighted. But she went, with their children, whose education she oversaw with a gimlet eye at the International School in Beijing.

The second was Rajendra Nath, a Frenchman of Indian birth. Raje, as he is called, had been a regular officer in the Indian Navy, in charge of engineering in the naval air service. When he decided to leave military service and was not allowed to resign from the navy, he ran for the Indian parliament and won a seat. The Indian constitution forbids one person to hold two government positions at the same time, and Raje was finally able to resign his navy commission. Later he worked for the International Civil Aeronautics Organization, a United Nations agency in Montreal, then for Airbus Industrie in Toulouse, and finally as an engineering section chief in SNECMA, the French aircraft engine company. He became a French citizen and married a Frenchwoman. SNECMA had sent him to Cincinnati, Ohio, as a liaison engineer with GE in our several cooperative programs, and Raje worked as part of the GE airline marketing team.

When professional people are sent on foreign assignments, the employer sometimes pays a premium salary to make an assignment attractive where the living and working conditions are unusually difficult. Typically, the employer "keeps them whole" financially against the additional expenses of living abroad. There is the cost of housing abroad in addition to maintaining the residence many of them own at home; the additional cost of schooling for children, sometimes boarding school if nothing is available in the work location; the higher cost of food and other articles of daily consumption; the cost of foreign taxes; the costs of a car, and perhaps a driver in countries such as China where foreigners may be wise not to drive themselves; and a host of other similar expenses. In some countries, language, customs, and social attitudes are such formidable barriers for "Europeans," a term used in Asia that includes North Americans, that it is good household management and good politics to take on live-in help—cook, maid, gardener, driver.

Americans do not think of their own country as a foreign location. SNECMA had its scheme for compensating its people for working in Cincinnati, Ohio. I do not recall the details but it was quite generous. However, such special compensation lasted for six years only, after which the employee either returned to France and his normal French salary or was presumed to have become acclimatized to living conditions abroad and also received only his normal French salary.

Raje was approaching the six-year point and was happy to take an overseas position with GE, with all the new overseas benefits this involved, rather than to have his salary revert to the French level. The negotiation was not easy. First we had to get SNECMA's consent because the two companies had agreed not to poach each other's people, which in this case posed no problem. Second, we had to define a set of overseas benefits for a French employee because Raje would theoretically retire in France and would have to keep up French social security and pension payments. That kind of problem gives corporate bureaucrats many hours of entertainment and is also a good test of the true global nature of a company. We managed to solve all the problems, and Raje and his family went off to Beijing, where he and his sons undertook an intense study of Mandarin. He became fluent enough to conduct business in Chinese and the boys later set up a trading company; so one can assume that their fluency in Mandarin was also adequate.

The third man was a mainland Chinese, Xiao Shao Cheng, an ex-CAAC engineer fluent in English whom we hired from Evergreen (years later, after a training assignment in the United States, he took over the Beijing office).

How did our international GE friends assess our new team? They grudgingly accepted Linton, because of the tour of duty he had put in with them in Pakistan and his obvious willingness to work with them. They liked Paul Sun—he was Chinese and fluent in Mandarin, two solid qualifications for doing business in China. They were less concerned than I was about his lack of experience in the aircraft engine business because they knew less about it. But they worried about Rajendra Nath: the Chinese have strong racial prejudices against Indians, said a British adviser of theirs in Hong Kong. The Chinese generally tend to consider themselves superior to all other races, but I never saw any effect of such prejudice against Nath and he said that he had not either. I doubt that GE lost any Chinese orders because Nath represented us.

The Zong Shan Affair

Marketing aircraft engines in China was a new task for Americans, and the challenge was always to sort out genuine opportunities from the false, to distinguish between real customers and con artists. Many propositions came to us from entrepreneurs who claimed to be connected to the top people in the State Council; we used to err on the side of caution and may have turned away some authentic sales opportunities. This account of our activities in China must necessarily be abbreviated; volumes could be written about our experiences. But there is one encounter included as a cautionary tale.

One day in 1988, Brian Rowe received a call from an old friend, the president of Rolls–Royce North America. They had been apprentices together at DeHavilland Engines in the early 1950s. He called to introduce Robert Gormley, a retired U.S. Navy rear admiral who acted as a consultant to Rolls–Royce in connection with the U.S. Navy's T-45 trainer program and was now involved in marketing American military equipment in southeast Asia.

Gormley told an interesting tale. His partner in Hong Kong, a respected Chinese businessman, had been approached by a daughter of Deng Xiaoping with the message that China wanted the GE F404 engine and that her father wanted him to arrange it. The F404 is the engine in the American F-18 fighter and a number of other advanced military aircraft in Sweden, India, and France, and internationally it was acknowledged as the best of its class.

It seemed odd that the message would come through such an informal channel. If the Chinese government were seriously interested, it could have said so to the U.S. government; directly to GE's office in Beijing; or to Washington, D.C., through the Chinese embassy. Why this approach through the underbrush?

This was the time during which the "princelings," the sons and daughters of important Chinese government officials, began to show up in business deals in Hong Kong and elsewhere. We had, for example, signed a sales representation agreement with a Chinese company called Polytechnology, Ltd., well connected with the PLA. The Polytechnology vice president with whom we signed was a young man in a black leather motorcycle jacket, with a thin mustache and a cigarette constantly in his mouth, the son of an important official—smart, self-confident, and quite effective in making spare parts sales for us. Perhaps the F404 deal was of this "princely" kind, perhaps not. Gormley did not know but he had confidence in his partner in Hong Kong and urged us to investigate. We informed the U.S. State Department that we were inclined to take a look. It was a sign of the times that we were not told to stop.

The first step was to meet Gormley's partner. Gormley and I flew to Hong Kong to meet with him in Abington's office. He was a dignified gentleman in his 60s, an American of Chinese birth who divided his time between offices in California and Hong Kong. He told us the same story we had heard from Gormley and vouched for the fact that Deng's putative relative was the source of the request. The General Secretary of the Chinese Communist Party had blessed his daughter's making a deal to get the GE F404 engine into China. We were puzzled: the F404 is a military engine for supersonic fighters. "For what purpose?" we asked. Our guest had no idea but suggested that it might be for police applications. We agreed among ourselves that he was doing no more than reporting that the daughter of the Chinese political leader had asked for an American product; beyond that, he understood nothing about the deal but clearly expected that he and Gormley would be in line for some kind of finder's fee.

We asked Gormley to arrange a meeting with the Chinese, which he said that Brian Rowe would be expected to attend; Rowe, fortunately, had another commitment, and Robert Turnbull, vice president of our Military Engine Projects Division, took his place. An experienced manager and engineer and a man of enormous enthusiasm, he was like all of us focused on expanding the business and welcomed opportunities to put one of our engine designs into a new application. There were plenty of other people around to make sure that all the legalities would be in place.

Turnbull came back from China bubbling. A large limousine had met him at the airport in Beijing and whisked him to a guest suite in the Diaoyutai compound. There had been a banquet in the Great Hall of the People, with a veteran PLA general and cameramen from the television news in attendance. And he had returned carrying a document, a memorandum of understanding that set out how our engine would enter the Chinese market. But the document presented certain anomalies that raised our suspicions. It was typed on plain white paper and described the mutual interest of Zong Shan Enterprise, Ltd., and GE to enter into a joint venture to produce the F404 in China, all the details to be worked out later. Turnbull had no clear idea who Zong Shan Enterprise was or what its capabilities and connections to a customer might be. Nor could Gormley shed any further light on the problem. We needed to know a lot more before we could further commit ourselves.

I asked Gormley to arrange a follow-up meeting. He would take along Frank Chang, his lawyer from Hong Kong, as an interpreter, and I would take Bob Linton and Paul Sun from our Beijing office. My charge from Turnbull and Rowe was to pursue the matter if it proved to be legitimate, otherwise to allow the memorandum of understanding to lapse. In Beijing, Gormley told us that the meeting with Zong Shan would take place in Harbin, a large city in Manchuria, because the municipal government planned to invest in the proposed joint venture. Linton and I immediately went to the Ministry of Aerospace Industry to meet with our friends from the aircraft engine department, the people who were the nominal owners of Liming and other Chinese aircraft engine factories. Who was Zong Shan Enterprise? we asked them. Was there a real opportunity for a GE military engine in a PLA aircraft? I kept remembering the Spey. With unexpected diffidence, they replied that in the new China we were obviously free to work with whomever we wished, but they wondered why we would want to want to go off on a wild goose chase like this. One of them reached into his desk and brought out a newspaper clipping. It reported, he said, that a notorious swindler had absconded with some money and the official seal of an enterprise in Harbin, and that the Bureau of Public Safety was looking for him. This was the man, he said, who was the organizer of Zong Shan Enterprise.

I reported this to Gormley and Frank Chang that evening. None of us was surprised. The most optimistic construction we could put on the information was that the Ministry might be jealous of a well-connected entrepreneur, but it was equally likely that the Zhong Shan Enterprise was fraudulent. There was only one way to find out. The next morning we went to Linton's office, where the Zong Shan people would pick us up for the trip to Harbin. They arrived at 11:00 a.m., half a dozen of them. Their leader announced that the site of the meeting had been changed from Harbin to Beidahe and that a bus was outside to take us there. We would stay at a hotel for several nights.

Beidahe is the Palm Beach of China, a resort on the coast of the Gulf of Bohai, about 300 km east of Beijing. During the hottest days of the summer, the bigwigs of the State Council leave the Zhongnanhai compound in Beijing and repair to the seaside, where they and their families swim and play tennis and relax. They also use the time to decide on policy for the party congresses that are convened in October.

We boarded the bus—Gormley, Frank Chang, Bob Linton, Paul Sun, and I, plus six Chinese presumably representing Zong Shan. Eight hours later, at sunset, we reached Beidahe and the bus driver found his way to the hotel near the shore where we were to stay. Another delegation from Zong Shan Enterprise was waiting for us and the obligatory banquet was all set to go the moment we could sit down. We checked into our rooms and then went to one of the dining rooms. It was late and we were tired. I remember little about the meal. We must have met the chief promoter but I have no recollection of it. I remember only walking back in the dark to the building where my room was, past a small lily pond. As I passed the pond, it burst into loud quacks—not from ducks, but bullfrogs that protested at being disturbed.

The next day we met with the Zong Shan people in a large conference room. There were elaborate introductions. Their party included an aged gentleman, a retired general who was a veteran of the Eighth Route Army with which Mao and Chou and Deng had retreated from Chiang Kai Shek's forces and then fought their way back into power. He was present, I supposed, for the same reason British and American companies have such military relics on their boards: to reflect some luster and suggest access to the leaders of the nation. The head of the Zong Shan Enterprise group was a thin nervous man, a chain smoker, unable to speak English. I, of course, am unable to speak Chinese. We depended on interpreters and that became an increasingly difficult business. Zong Shan had prepared a document of many pages, in Chinese and English, which they proposed that we discuss and then sign to launch our joint venture. It was slow going and we had great trouble understanding the drift of their proposal.

We were invited by the provincial government to see for ourselves what it was doing to stimulate economic development. We visited the coal port of Chinghuandao, large and quite modern, from which China exports coal all over the Pacific and even to Europe. We also visited a factory making steel tubing for automotive and appliance applications. This was a small modern enterprise, a joint venture with the Australian subsidiary of Bundy Steel Tubing, Ltd. The production machinery had been brought in from Australia and the copper-plated steel input material from Belgium and Italy. The provincial government had built the factory and the access road and utilities, and provided the labor force. The general manager and his process engineer, both Australians, were the only two foreigners there, both on one-year contracts. They did no marketing, he told me. Customers came to the loading dock to plead for tubing. All they had to do was to make it. Productivity of his Chinese employees was higher than that in Australia, he said with a little embarrassment.

That evening, we gave our obligatory return banquet in a restaurant not far from our hotel. About 100 people from Zong Shan and the provincial administration attended. A banner greeted us announcing the signing of the joint venture, and a man in a tuxedo introduced himself as the hotel's food-and-beverage manager and hoped that everything would be to our satisfaction. It was an elaborate meal, involving a lot of drinking and many toasts. Some of the drinking, it became clear, was competitive: young men on the Zong Shan side, invited for reasons not apparent at the time, came to toast us one by one, challenging us to empty our glasses. Photographers appeared and we were invited to have our picture taken signing a document in a big red folder in front of the banner. I announced that

I would be happy to be photographed, but not in front of a banner announcing that we had signed a joint venture agreement. That was premature. The provincial people were disappointed and the Zong Shan group showed a little annoyance. There were no pictures.

"Chairman Bob," Admiral Gormley said to me, "our side is taking casualties." At a table near ours, one of our people had slumped forward, his face in his dinner plate. Two helpers carried him out and put him in a car to take him back to the hotel. A band played loud Western music and dancing began. The only women present were a couple of interpreters and the waitresses, who were all dressed in snappy tuxedos. The old general sat at our head table and invited me to dance. I joined him, the only time so far in my life that I have danced with a general. He was in his 80s, like me more than a little drunk, and followed quite well. One of the young drinkers invited me to dance. I thanked him and declined. He took my arm to urge me to the dance floor. An interpreter said something, presumably that I was a bad sport, and the young man went away. As the nominal host, I stood up, thanked everyone for coming and announced that the party was over. Bob Linton went to pay the bill—the single most expensive meal, banquet or otherwise, that I have ever attended, about $10,000. Luckily, they accepted Linton's American Express credit card.

The next morning, we decided that we had to bring this affair to an end. In the big conference room, I proposed a small meeting between the chief promoter, his interpreter, Admiral Gormley, Frank Chang, and myself. We met in my hotel room and I asked our host what he really wanted. After some discussion, that became clear: he wanted to sign an agreement with us, stating that the province would put up a factory building and supply the workers. GE would bring in machinery and tooling, materials and technical data, and manufacture the F404 engine. For what market, I asked. For the Chinese market and for export, he replied. The draft document he had given us described just such an arrangement, he said.

It was not impossible, I thought, that someone in the State Council had said to one of his pals, "We could use the F404 engine. Why don't you see if GE will set up a joint venture factory here to do that, and to export some?" It was the pattern followed by Nike for athletic shoes and other American and European manufacturers, but not until now for capital goods such as aircraft engines. I responded that GE was already serving the world market for the F404 quite well from our own factory. Our interest was only in the Chinese market, and if China needed only a small number of F404 engines, importing them from the United States was the best and cheapest solution. If China needed a large number, more than say 300, and were prepared to pay the cost of setting up a production line, GE would consider a technical assistance and license agreement to have the engines assembled in China and some of their parts made there. Such an agreement would have to be approved by the governments of China and the United States. But I could not sign the Zhong Shan document. We were prepared to write down something that described what I had just told him, which GE could sign. Even that, I added, was contingent on evidence that the Chinese government as the ultimate customer was serious about the program, intended to use the F404 engine, and endorsed the concept of having it made in China. What would satisfy GE, I added, was a notification to this effect from the Secretary of the Defense Committee of the State Council, which Deng Xiaoping chaired.

Our friend sucked on his cigarette and said he would have to think about that. He would let us know that night if he could get us such a letter. How about a signature from the old general? That would not suffice, I told him. If he could not assure us that we would get such a letter, we would have to return to Beijing. We parted company and returned to our rooms. We heard no more from him and had dinner by ourselves. We decided to leave. Like Zong Shan Enterprise, the bus that brought us to Beidahe had vanished. The train to Beijing would leave at 6 a.m. We were up at 4 a.m. to pay our bills and get to the station.

The train was packed and stopped often. The ride must have lasted almost 8 hours. We kept going over the bizarre events of the previous three days, what we might have done, and what we should do now. Linton told us about a strange young man in a green double-breasted suit who delivered the first messages in Beijing and then had shown up briefly at the hotel in Beidahe, without ever joining the discussions there. A green suit is hard to overlook. Who was he and whom did he represent? Gormley and Chang were disappointed that nothing had come of the adventure but also embarrassed that they had dragged us into it. There was nothing to be concerned about, I told them. It might have turned out to be great: nothing ventured, nothing gained.

Had we been in the business of making and selling cookware, hair dryers, telephone handsets, this bizarre episode would have been quite normal. The Chinese paradigm for international trade is, "Have foreigners make their products in China to sell on the world market." Perhaps Deng Xiaoping really did say to his daughter, "See if you can get GE to do that with the F404. We would even take some for our own air force." And perhaps the thin man, and not the bureaucrats of CATIC, really was the embodiment of Deng's new economic policy: China will put up the factories and bring in the workers and the foreigners will bring in capital, technology, and machinery, and manufacture products in China for sale all over the world.

Back in Beijing, Linton told me that he wanted me to meet another unusual character, not for any specific purpose but with the possibility that he might be able to shed light on the affair. This was Sidney Ruttenberg, an American who had served in the U.S. Army in China in 1949. Evidently with strong sympathy for the Chinese Communists, Ruttenberg had stayed in China when the American troops left. Somewhere along the way he had been arrested and imprisoned as an alleged Central Intelligence Agency agent. His cell-mate, he said, had been Deng Xiaoping. There are many people who claim that they have been in prison with Deng. In any case, Ruttenberg now runs a Beijing import–export firm on the strength of his contacts and knowledge of the Chinese power structure. Linton and I went to visit him. Ruttenberg listened to our account of the Beidahe discussions and told us that for $15,000 and expenses he could find out for us what the truth was. I reckoned that we could find out for less.

Linton then arranged for a meeting with the vice chief of staff of the PLA air force. General Win turned out to be a tall man with bright blue eyes and a good knowledge of English. I decided on the direct approach and asked him whether the PLA air force wanted the F404 engine. Yes, he said, we want to put the engine in our F-7 fighter, the Chinese version of the MiG 21. How many? Let us say 50 aircraft, he replied; perhaps also in the future in the F-8, a larger twin-engined fighter. How might such a program be accomplished? I answered that the quickest and cheapest way would be to import the engines from GE. Only for a much larger number, say more than 300, would it make sense to assemble engines in China. I recommended that an American aircraft manufacturer be involved to help design the installation and the flight test, and recommended Grumman, which had worked with GE on a similar program in Singapore. If General Win wanted to go ahead, I recommended that he lend us one of his staff officers to prepare a plan for his approval. Although he did not agree on the spot to go ahead, the technical discussions between the Chinese air force, Grumman, and GE continued for some months. Then a major political development put a halt to all such matters.

Tienanmen

The easing of China's relations with the West and the new economic policies inevitably led to calls for democratic reforms. A few in the State Council appeared sympathetic, but most of the Politburo was not prepared to accept the slightest challenge to the Communist Party's political hegemony. In 1989, there were more and more demonstrations in Beijing, led by university students. Ominously, they were joined by some factory workers. The situation became more tense by the day, and we decided to pull our people out of China. It was early June, and the biennial Paris air show was about to take place: the great international convention and trade fair of the industry. I asked Linton to make sure that his people went to the United States on home leave. He and Raje Nath would join us in Paris to take care of any Chinese visitors who might show up.

Just before the government crackdown that sent PLA tanks against the student demonstrators, one of Raje's sons went to Tienanmen Square with a videocamera. He climbed up one of the ceremonial lamp posts to get a good view. When the police moved against the students, they also grabbed him, pulled him down from the lamp pole, and arrested him, ostensibly for breaking one of the glass globes and more likely for taking pictures of the events. Raje raced to the police station, where he explained that this had all been a terrible misunderstanding, that he and his son apologized humbly for breaking the glass globe and disturbing the peace of the Chinese scene. They let him go.

I was in Paris, working on the arrangements for the GE exhibit and chalet at the air show. Every evening, CNN showed terrible scenes of tanks moving against the students and Chinese policemen frog-marching arrested students into custody. Two militiamen held each handcuffed person by the elbows and forced his head forward and down, holding him by the hair.

The official Chinese delegation arrived in Paris. Jean Billien, the president of CFMI, arranged to meet them for breakfast at our hotel. A dozen Chinese showed up, all of them well known to us from many previous meetings in Beijing. All were wearing Western clothes except vice minister He of the Ministry of Aviation Industry, who was dressed in a dark gray Chinese uniform. There was little small talk. The vice minister began to harangue us, particularly me, about American effrontery in encouraging Chinese students to create political disturbances, and for condemning the Chinese government's stern actions to restore order. Such brazen interference in another country's internal affairs, he said, could only hurt the development of friendly commercial relations. I responded that we were a private company, and that he should address his remarks to our government. He would have none of that. "You know General Haig?" he demanded, referring to the former U.S. Secretary of State who was now an advisor to Pratt & Whitney in China. Haig had been the first, and certainly the most prominent, American to appear in public with the leaders of the Chinese state after the Tienanmen massacre, drinking a toast to something or other. It was not a performance I would have wished from a prominent American. "No doubt your company has someone like Mr. Haig," the vice minister continued, "who can pass this message to your president." He rose. His entourage, with silent and perhaps embarrassed faces, rose and left with him.

Jean Billien was appalled. Our business was held hostage to political tensions between the Chinese and American governments. Should we write a letter to the White House? If relations between the countries were to be strained by the events in China, I thought, the impact on CFMI's market opportunities would be unlikely to change the mind of the U.S. government. Nor would it be seemly or well-received at home for a company to complain about its business prospects while the whole world was watching the brutal treatment meted out to the protesters in China. I suggested that Billien inform his government, the legal owner of SNECMA, while I went to the U.S. Embassy to meet with representatives of the U.S. Departments of Commerce and State to report the substance of Vice Minister He's harangue.

Relations with China remained strained for some months, except for General Haig's toast. Despite that, we saw little effect on the conduct of business, either from the Chinese side or that of our governments, with one important exception: all discussions concerning the F404 military engine for the Chinese F-7 fighter were stopped. We were informed that our licenses to export data of any kind were withdrawn and that further discussions of the program with the Chinese would be against U.S. interests. Grumman received the same instructions.

One traditional Chinese curio consists of a carved ivory ball, inside which is another carved ivory ball, separated from the outside and free to rotate. Inside that can be a third, also carved but this time not hollow, a monument to the carver's manual dexterity and fine discrimination. Political relations between the United States and China can be like that. Nothing is simple and direct, and every act has consequences, some intended and some not. The U.S. government had approved the development of an engine for the Taiwanese IDF on the rationale that the basic engine was commercially certified for business jets and, therefore, did not come under the limits on military exports imposed by the Shanghai Communiqué between the United States and the mainland. In much the same way, the U.S. government approved the sale of Sikorsky S-70 helicopters to the PLA in the mid-1980s.

The S-70 is a renamed BlackHawk helicopter, the basic utility helicopter of the U.S. armed forces, with a civil model number. Like cargo aircraft, these machines are in essence flying trucks. Paint them green and they are military; shine up the aluminum and they are civilian. The PLA needed modern utility helicopters and held a competition between Aerospatiale's Super Puma, Bell Helicopter's 214ST, and the Sikorsky S-70, the final test to be a high-altitude flight evaluation in Tibet, somewhere between Lhasa and the Indian frontier. The Bell and Sikorsky helicopters both use GE T700 engines, and we sent a couple of field service engineers to support the tests. The S-70 proved to be the most capable of the three helicopters, capable of lifting greater loads and with better maneuverability than the others at such elevations. The Chinese ordered 20 from Sikorsky, which renamed it the PeaceHawk for the sake of making it different from the U.S. military model. The U.S. government approved the sale and the helicopters were delivered over a couple of years. Sikorsky and GE sent field service engineers to support operation and maintenance of the birds. Paul Therrien was in charge of them as well as the technical representatives GE sent later to support operation of our commercial engines. Their view of disciplined helicopter operations and that of the PLA diverged somewhat. Our people were startled to see Chinese pilots arrive at a base, get into the helicopter, turn on the ignition, and fly off with no preflight inspection or instruction about operating limits and precautions. Maintenance was not up to our standards either. Our people scrambled to keep a minimum number of the helicopters in safe flying condition.

When all U.S. export licenses for military equipment were withdrawn after the Tiananmen massacre in June 1989 that included the T700 engines in the Sikorsky helicopters. We were not allowed to ship spare parts to China nor to provide service bulletins or the latest information about engine problems elsewhere in the world. Although many of the helicopters sat on the ground because we had declared them not airworthy, we had little confidence that the PLA would comply with our recommendations. There was nothing we could do to stop them from flying their helicopters if a pilot took it into his head to do so or was ordered to do so. We told the State Department that in such circumstances there could be a deterioration of flight safety, in the worst case leading to an accident. The PLA used the helicopters all over China on missions that were often as much civilian as military. We did not want the United States or ourselves blamed for an accident attributable to a lack of product support after the export licenses had been suspended. It was a justifiable concern, but the political revulsion over China's behavior was so strong that the suspension of export licenses stayed in effect for quite some time. More than one year elapsed before we were allowed to resume deliveries of spare parts.

I saw Vice Minister He once more after that Paris breakfast meeting, in Singapore in January 1990. We were all there for the Singapore air show, the Asian response to Paris. Billien, Linton, and I again invited He and his party to dinner. He came, brought a big bouquet of flowers for my wife, and bubbled with charm and enthusiasm. Billien and He talked about business as if no shadow had ever passed between us. That, too, was probably a message to the French and American governments.

Winding It Up

Face and title are important in China. Pratt & Whitney, our biggest competitor for airline engines, had appointed a senior man as president of its Chinese programs. He was based at the corporate headquarters in Hartford with access to Pratt & Whitney's top management. With his gray hair and ability to make quick commitments without having to call back home for approval, he made a very positive impression on the airline people in China. With this example in mind, I proposed to Brian Rowe that I call myself the president of GE Aircraft Engines—China. The vice president of airline marketing, who was also present, was scornful. The Chinese know the difference, he said, between a real president and a fake one. "OK," Rowe said to him, "You can have the job." He was instantly glad to have me inherit the position.

For the next couple of years I visited our customers in China at least three times a year, to support the work of the marketing and technical people stationed in China and to demonstrate the importance GE ascribed to its Chinese operations. Before I retired in 1992, I made a farewell round of our business contacts. The trip took me to Shenyang once more, to Xi'an, and of course to all our friends in Beijing. There, with the China State Shipbuilding Corporation, my hosts indulged themselves and me in what they considered one of life's great luxuries. I was given the dubious pleasure of having to eat a piece of camel hump.

I left China, and a few months later retired after 42 years and 3 months working at GE in the aircraft engine business. Linton was transferred shortly afterward to straighten out our program in Israel, where we had become involved in a tragic criminal affair (mentioned briefly in Chapter 8) in which one of our people and an Israeli air force officer had conspired to skim large sums of money from our contract to deliver engines and equipment to the Israeli air force. Raje Nath was transferred to India to add muscle to our marketing activity there. Paul Sun returned to Cincinnati to work on technical assistance programs in Taiwan, China, Indonesia, and Australia. And Xiao Shaocheng, whom we had hired from CAAC, was sent to Cincinnati with his family for one year to work inside the Airline Marketing Division. There he would become familiar with what we were doing and how, and our sales people would get to know him and become comfortable working with him when he returned to China, so that we could avoid a misstep like some of those in the past.

The man who took over from Linton in China was David Voeller, whom I had hired for that exact purpose eight years before. Voeller, an engineer from Oklahoma, had gone to China for a company called Unit Rig, a manufacturer of giant trucks used in open pit mines, trucks capable of carrying 150 tons of ore. Unit Rig had sold two such trucks for a big open pit coal mine developed by Occidental Petroleum, with the understanding that some truck parts would be made in China. Then in his early 20s, Voeller packed catalog cases with drawings and specifications and off he went. He found Chinese factories to make the parts, Chinese steel mills to furnish material that met the specifications, and he supervised the manufacture and inspection of the parts. Then the Occidental Petroleum venture fell on hard times. Inflation was raging, and the Chinese government cut off all funding to grandiose schemes such as this coal mine. Unit Rig called Dave back to Tulsa, Oklahoma. Unwilling to sit around until China's economy recovered, he enrolled at Columbia University in a graduate program of Asian studies. One of his fellow students was the son of James Krebs, our vice president of Military Engine Programs. Krebs met Voeller, liked him, and suggested that I interview him.

I have interviewed many salesmen over the years, but few engineers with that kind of background, that kind of self-sufficiency and enterprise. Voeller could speak passable Chinese and in Beijing had met a young Chinese woman from Hong Kong working for Citibank, whom he married, a little to her parents' dismay because he was a red-haired American. I hired him as a possible replacement for Walter Chang. He worked in my operation for a while and then in the Airline Marketing Division to manage the support of sales campaigns and participate in them, beginning with Evergreen Airlines in Taiwan, a new enterprise getting itself into position to carry traffic directly between China and Taiwan as soon as politics permitted. Voeller finally went to China with the enthusiastic endorsement of Jack Welch, the GE chairman, to take over and expand the GE Aircraft Engines activities there. When he was promoted into his next assignment, Xiao Shaocheng was ready to take over from him. That is not a bad plan of management succession. GE is fortunate to have the resources to make it happen.

What was the benefit of all of these adventures for GE? The office in Beijing cost about $1–1.5 million/year; GE's airline salesmen spent perhaps the same amount again. In 1979, when we began our market development work in China, GE's share of the aircraft engine market there was zero. Most of the engines used by CAAC were Soviet or of Soviet design built in China under license. The CAAC had still a few of the Tridents it had bought from Pakistan, with Rolls–Royce Spey engines. The important new aircraft acquisitions were from Boeing, 707 airliners with Pratt & Whitney JT3D engines. Fifteen years later, GE and CFMI's combined share of the rapidly growing Chinese commercial engine market was between 35 and 50%, with annual sales fluctuating between $300 and $600 million. That may not be market dominance, but it is still a considerable achievement.

Chapter 17

Developing the Russian Market

At the end of World War II, the United States and the Soviet Union competed in hiring German scientists and engineers who had been involved in rocket and jet propulsion. A group of German engine designers was taken to France, where they designed the Atar series of engines at the predecessor company of SNECMA. Of those who were brought to the United States, some went to Wright–Patterson Air Force Base, the headquarters of Air Force Systems Command, and some to Tullahoma, Tennessee, to build the giant altitude test cell at the Arnold Engineering Center—an enlarged and improved version of the test cell the same engineers had built in Bavaria for BMW—and some to Lycoming, in Stratford, Connecticut, to design the engines that became the T53 and T55 turboshafts. Wernher Von Braun and some of his team from Peenemünde went to Huntsville, Alabama, to work in the U.S. Army's ballistic missile program. Life in the United States and France was better in 1945 and 1946 than it was in Germany, so that there was no great degree of coercion involved in what was code-named "Operation Paper Clip." The Soviets were less gentle. They scooped up the entire Junkers engineering staff and set them down in an isolated base in Siberia to design a 12,000-hp turboprop engine.

In the late 1940s, the Soviet Union bought 50 Nene turbojet engines from Rolls–Royce. The engines in the MiG 15 fighters that went into action against the United States in Korea in 1950 were Soviet derivatives of the Nene. One big difference between the Soviet Union and China was that the USSR invested heavily since 1950 in research and development of aircraft gas turbine engines, and had set up several large factories for their manufacture.

During the Cold War, the United States and the Soviet Union each built up a large and self-sufficient aircraft engine industry to satisfy its peak military demand. Neither expected to compete in the other's commercial market and yet, in 1976, GE received a request from the Soviet embassy in Washington, D.C., to quote on the delivery of 12 CF6-50 airline engines. A similar request went to Pratt & Whitney for the JT9D. A Soviet government delegation received U.S. government approval to visit Pratt & Whitney's factory in East Hartford, Connecticut, and GE's in Evendale, Ohio. The visits took place, certainly to GE's surprise. The U.S. State Department advised GE to limit all displays and presentations to commercial engines and to tell the Soviet visitors no more than GE would be prepared to show or tell a visitor from a real competitor such as Pratt & Whitney. Curiously, no GE vice president other than the head of manufacturing found himself able to be at the factory during the visit—a fact not lost on the minister leading the Soviet delegation. GE's Cold Warrior factory workers had to be cautioned to act with politeness toward the visitors and to refrain from overt displays of hostility.

Following the visit by the Soviets, who went also to Boeing and Douglas, the U.S. Commerce Department gave GE an export license for technical brochures supporting further discussions in Moscow. Brian Rowe, then head of the Airline Programs Division, led a small GE group that met with academician Vladimir Sosunov, the deputy chief of propulsion research at TsAGI, the central institute for aeronautical and hydrodynamic research, the Russian counterpart to NASA's aeronautical activities. While they were there, the U.S. House of Representatives sent a resolution signed by some 80 Congressmen to the White House, condemning in the strongest terms any notion of selling American aircraft engines to the Soviets. GE concluded that it had much more to lose in Washington than it could gain in Moscow and returned its export license to the U.S. Commerce Department with thanks.

A few weeks later, the Soviet government trading company Amtorg asked GE's International Licensing Department whether GE would sell a license for a patent on turbine blade cooling. The patent is of course available to anyone, including Amtorg, who pays the small fee at the U.S. Patent Office. What Amtorg was asking for was the right to make turbine blades without being sued by GE for patent infringement. Despite the fact that no technology transfer was involved, GE said no because it wanted to avoid any public controversy in the United States about dealing with the Soviets on aircraft engines.

In 1989, Mikhail Gorbachev proclaimed *glasnost* and *perestroika*. The Berlin Wall came down, and relations between the USSR and the United States changed dramatically. The U.S.–Soviet joint chamber of commerce sponsored a big trade exhibition in Moscow. GE had a prominent exhibit, as did other big firms such as IBM, Procter & Gamble, and Pepsi Cola. Gorbachev visited the IBM and GE exhibits, where he talked briefly with GE executive vice president Paul Van Orden. Soon, Professor Sosunov of TsAGI followed to send his best regards to Brian Rowe and to invite GE to hold a technical seminar in Moscow about its commercial engines. During the exhibition, GE gave a reception for invited Soviet guests, officials of the ministries, and agencies with whom we hoped to do business. Any reception with a buffet and drinks was sure to draw a crowd, our Russian experts told us; whether the buffet was the lure or not, we were able to entertain Professor Sosunov and academician Gendrikh Novozhilov, the head of the Ilyushin aircraft design bureau. Both repeated the invitation to hold a technical seminar in Moscow during an international aircraft propulsion conference planned for 1990.

We accepted the Soviet invitation, as did many other American and European companies. This time, GE Aircraft Engines sent a large team and a sizeable exhibit, including a full-scale CFM56 mockup. SNECMA had its exhibit nearby and brought along a chef and food—the only source of lunch in the entire exhibit hall as far as we could tell. Our exhibit was visited by many Soviet engineers and managers, with whom we managed to exchange business cards and initiate exploratory discussions. Some were romantics, such as three sincere veterans of the industry who proposed that GE develop a dual propulsion system for a complicated high-altitude transport of global range, with separate engines for takeoff and climb, and for high-altitude cruise. Others were more practical. For example, would the GE T700 turboshaft engine fit into the next Mil helicopter? Mil had built more than 10,000 Mil 8 helicopters, many exported to Africa and South America. The new helicopter would doubtless be built in the same numbers, the commercial director of Mil told us. Two engines in each of 10,000 helicopters, plus the same number of spare engines—the numbers would make your head spin! The opportunity was so great that GE ought to set up a factory in Russia to satisfy this market. How much had Mil received for these exports, we asked. The commercial director did not know. Mil had transferred the completed aircraft to Aviaexport, a state trading company, which made the actual sale. Was the business profitable for Mil? The commercial director smiled. "I received my salary," he answered. This discussion typified the discontinuity of experience between the Soviet industry and ours. They talked of production numbers far bigger than we estimated the market could sustain and their view of cost and price was cavalier. They were not trying to mislead us but were addressing a potential market in light of what they knew, just as we were. They were quite sure of two things—their technical capability was at least as good as ours and their costs certainly were far lower.

Most of the exhibitors stayed at the Kosmos Hotel, a large commercial hotel next to the park in which the exhibition hall was located. GE maintained a "hospitality suite" in the hotel to which we could invite customers for discussions after the exhibition, and we also booked a large auditorium in the hotel for our technology symposium. The GE corporate office in Moscow had sent out invitations to the government and to the Soviet aerospace firms to attend presentations on GE's large and small commercial engines and their marine and industrial derivative gas turbines. As one of the world's largest producers of oil and natural gas, the former USSR is a lively market for industrial gas turbines. The small size, light weight, and high power of industrial gas turbines derived from aircraft engines make them very attractive for installation in remote pipeline pumping stations.

About 300 Soviet scientists, engineers, academics, and administrators came to the conference, no doubt accompanied by several technical intelligence specialists from the Soviet air force. They were probably curious about the level of our technology compared to theirs. But the burning issue was money: for years an industry as large and as wide-ranging as that in the United States had been nourished by a stream of development funding from the Soviet government and by large production orders. *Perestroika* and the evident bankruptcy of the Soviet government put an end to that. In the United States, a similar phenomenon was called the "peace dividend." In the former Soviet Union, tens of thousands of technocrats confident of their own technical capability saw nothing in their future other than the closure of design bureaus and factories and mass unemployment. We had invited questions from the audience. The most important was "Where does the money come from for new programs?"

The hall fell silent. I do not remember precisely how I answered at the time. But I have thought hard about that question and have told the Russians, then and since, that there are three sources: 1) the customer, paying separately for research and development or paying for it as part of the purchase price for engines well in advance of their development and delivery; 2) external investors in expectation of future profits; or 3) profits retained from previous programs.

The first option was one the Soviets understood very well: that is how their government paid for the development of their military engines and most of their commercial engines (it is also how the U.S. government pays for the development of military engines). For the second, our guests knew little about external investors, although they had a vague notion that Europe and the United States may be a source of development money, replacing the dried-up flow of funds from the Russian ministries. And regarding previous profits, from what we had been hearing our visitors had little understanding of the meaning of profits and how they come about. Nor did they have any retained earnings large enough to be a source of program funding.

When the conference adjourned for lunch, two men approached. They represented the Ministry of Machine Building. Would I be prepared to meet with the minister? I agreed to do so. A couple of days later, the two men picked me up at the hotel, with Michael Elman, the GE engineer who served as our chief interpreter, and drove us to the headquarters of the Ministry in a compound of old red brick industrial buildings and laboratories somewhere in Moscow. The Ministry, or that part of it we saw, was responsible for space propulsion and space power, including the "Energiya" rocket engine. We were taken on a tour of its test facilities and shown remarkable devices: a powerful electron beam generator, plasma arcs, and all kinds of other spectacular apparatus. What they wanted, the minister and his deputy told us, was American customers for their expertise. The minister was emphatic: they had first-class technology, they had 4000 scientists on their payroll, they were getting little money from the Soviet state, and they needed funding to keep their projects going. And to keep paying their staff, the deputy added. I promised to carry the message to the United States.

GE was no longer in the space power business, but Pratt & Whitney and other companies were, on contract to NASA and the U.S. DOD. Back in the United States, I reported what we had heard, and five years later, as this is being written, there is a lively cooperation between the Russians and American contractors to use the Russian RD-180 rocket engine for satellite launches.

Out of such beginnings, we tried to crystallize a GE strategy for the USSR and Russia after the Soviet Union was broken into its parts at the end of 1991. GE wanted to sell its engines into this last unserved market but, as always, there was strong competition, this time not from Pratt & Whitney or Rolls–Royce, although they had the same objective as GE; the real competition came from the Soviet aircraft engine industry, which considered itself just as capable as that of the West and saw no reason why it should concede any shred of what had been its protected home market to foreign carpetbaggers.

The Soviet engine industry supported itself while the USSR was walled off from the rest of the world, but everything had changed. Already in 1990, the Soviet and the soon-to-be Russian airlines saw a different world coming. Aeroflot and all the regional airlines into which it was broken up were flying to destinations outside the old Soviet borders where they could see that foreign airlines preferred American and European airliners—the airlines for lower operating costs and greater productivity, the passengers for the greater comfort and higher standard of reservation and cabin service. Young managers in the Russian airlines quickly understood that they had to upgrade their equipment and service to remain competitive.

LOT in Poland, Balkanair in Bulgaria, and CSA in Czechoslovakia quickly bought or leased airliners from Boeing. Balkanair was an interesting case. It was a much larger airline than one would expect to find in Bulgaria. Much of its business was flying tours from the former East Germany to the beaches on the Black Sea. Prices for such tours were low, and the airline was able to maintain this business even after the unification of East and West Germany. Balkanair bought Boeing 737 airliners and demonstrated quickly that their productivity was four or five times as great as that of the Tupolev 154s it had been using. The experience of LOT and CSA confirmed that. We began to understand why the Soviet manufacturing industry was so much bigger than our assessment of their market had indicated: it had been compensating for lower aircraft productivity by building four or five times as many airliners as ours to serve the market. Our yardstick had simply been wrong.

Aeroflot International arranged with Airbus Industrie and GE to lease five A310 airliners with GE CF6-80C engines. We got together with them at Tempelhof Airport in East Berlin, a convenient location for the Russians, so that our logistics people and theirs could discuss product support. Aeroflot asked how many spare engines they should order. Our people said, "Five aircraft: 10 installed engines; 15% spare engines, but since it'll be a small fleet, order two, or perhaps even three spare engines." The Russians could not believe their ears. For their Ilyushin 62 airliners they had 400% spare engines, four for every one installed in an aircraft. Our people were equally astounded. The Russians explained that one spare engine was stored at Tempelhof, another at Sheremetyovo airport in Moscow, so that the aircraft could without fail takeoff on its daily routes without being grounded for engine failure. A third engine would be in overhaul in the factory in Perm, 1000 km to the east of Moscow, and a fourth in transit to or from the factory for repair or overhaul. In fact, this is a reasonable way to maintain a high level of reliability and safety, as long as inventory is free, as long as there is no interest cost for working capital tied up in an inventory of spare engines. Marxist economics make it a reasonable solution.

In the West, money tied up in nonproductive equipment becomes an expensive operating loss. Americans and Europeans design their aircraft to be capable of revenue service as much as 10–12 hours/day—as much as 16 hours/day in some extreme cases. Any airframe or engine problem that keeps an aircraft sitting on the ground immobilizes a multimillion dollar investment. Interest expenses, depreciation, and lease charges do not stop. We design our systems so that a faulty component can be identified and removed from the aircraft on the flight line, replaced with another functioning one, releasing the aircraft for return to revenue service as quickly as possible. The faulty component is the only part sent to repair and overhaul.

That philosophical consideration, that the airline must pay interest expense on capital employed in the enterprise, affects the mechanical design of the engine. We have designed ours so that components can be removed and replaced easily by a few mechanics on the flight line. The major subassemblies are "modular" and can be separated from the rest of the engine without destroying the structural integrity of either. The Russians say that their design is also modular, but we saw that many of their flanges, for example, were much thinner than ours. That is efficient from the point of view of weight, but thin flanges are easily deformed in disassembly, particularly by inexperienced mechanics in the field. The Russian engines had not been designed for maintenance in the field but to be disassembled in the factory, where rigid tooling was available to keep the parts from being distorted.

This design philosophy also explains the huge size of the Soviet aircraft engine factories and the massive scale of their production. If their practice has been to build, let us say, four spare engines for every one installed in an aircraft and if installed engines come back to the factory at intervals of 1000 to 2000 hours, in contrast with the 5000 to 10,000 hours not unusual for Western engines, then it is no wonder that the typical Soviet engine model has been produced in numbers five times as great as ours. When Mil Helicopters' commercial manager forecast a market of 40,000 engines, he was merely reflecting conventional Soviet design practice. If we deflated his helicopter forecast by a factor of five to take into account the difference in productivity we had observed between the Boeing 737 and the Tupolev 154 airliners, and deflated the number of spare engines to a level of 15–25% of the number of installed engines according to our own experience, then the helicopter engine market would be 4800 over 20 years rather than 40,000. To us a production program of 4800 engines over 20 years, an average of 20 engines produced per month, would still be a very big program, well worth pursuing.

We went through similar calculations for the Soviet market for airliner engines. Because of the long distances between major cities and the relatively underdeveloped road and rail networks, passenger and freight traffic by air plays a larger role in the Russian Federation than in the United States and much larger than in Western Europe. Fuel costs to the airlines had been set by the state at levels far below what airlines in the rest of the world had to pay; so there had been no great emphasis on low specific fuel consumption. Fares had also been set very low to reflect the availability of air travel as a social good. Because demand for air travel is price-elastic, low fares meant full airplanes. At a conference in Washington on April 10, 1991, the chief engineer of the Progress Engine Design Bureau said that the fare from Zaporozhye in Ukraine to Moscow was 20 rubles, officially about $20, the same as a train ticket. The distance is about 600 miles. In March 1992, a ticket from Perm to Moscow, also 600 miles, cost a Russian $20. Foreigners paid $200. The flight was oversold and three people stood in the aisle of the Tupolev 154 without seats during takeoff. Fares were low enough that peddlers could fly from Tblisi, the capital of Georgia, to Moscow with a suitcase full of fresh fruit, sell it on the street market, and return to Georgia with a good profit. Russia is able to export crude oil to earn foreign currency. If the price of jet fuel in Russia were to rise to world prices to take the export opportunity into account, the cost of air travel would have to rise; higher costs would lead to higher ticket prices and, in turn, depressed demand. How then to estimate the size of this market for modern aircraft engines?

Based on all of the considerations of improved productivity and lower spare engine inventories, and on data from Boeing and GE's Airline Marketing Division, we forecast a passenger airline market of

2850 installed engines of the CFM56 class over 15 years, plus another 460 spare engines. Cargo aircraft such as the Ilyushin 76 could use another 400 installed and 100 spare engines in the same time period. We were looking at a potential market for large airliner engines of just under 4000 over 15 years, an average production of about 20 per month. Small jet airliners with 50 to 100 seats would require 900 smaller engines over 15 years, about five per month. Smaller turboprop airliners would require another five engines per month, a total of 1000.

To the Soviet engine experts, these estimates appeared wildly unrealistic. They were so much smaller than their production capacity or any numbers in their long-range plans that they could only conclude that we were trying to deceive them in some way. If our numbers were correct, the Soviet engine industry would have to contract to 10–20% of its current size (by no coincidence, we estimated the actual work load in a number of Soviet factories at about 10–20% of peak capacity), throwing hundreds of thousands of managers, engineers, and skilled workers out of work. What we said was simply not acceptable. Yet the numbers we forecast were big enough by our standards to make an attractive market, particularly if it were incremental to what we were already producing in the West. The challenge was to find an effective strategy to serve this market, in competition with the Soviet industry, keeping in mind the extreme shortage of foreign currency available to the airlines. We were not concerned about potential competition from the Soviets in Western markets where we felt that our products would be preferred because of their better fuel economy and dependable product support. The Soviets may offer aircraft at cutthroat prices, but such a strategy would very quickly be self-defeating.

The product support factor is very important. Soviet factories in a command economy were judged by their production output of complete engines. Their buyer was the Ministry of Aviation Industry, which would direct delivery to some airframe factory. There was very little direct communication between the factory and the operator, airline or air force squadron. Therefore, the manufacture of spare parts was a minor and relatively unimportant part of the entire production process. Two examples will illustrate the problem.

- The Indian navy used Soviet gas turbines to drive some of its destroyers. The navy could order spare parts for these only once a year, during a precise two-week period. If that was not done, the navy had to wait for another year to place the order.
- When we proposed to put CFM56 engines on Aeroflot International's Ilyushin 86 airliners, the airline's chief engineer told us that he would never agree. He had no concern about our engines or product support, but "Who," he asked, "will give me spare wheels and brakes, or windshield wipers and cockpit instruments? I cannot depend on Ilyushin for replacement parts."

GE decided on several strategies. The first was to sell engines installed in American or European aircraft to the Soviet airlines that flew international routes. Boeing and Airbus Industrie were marketing their airliners aggressively, and a market also slowly developed for used DC-10 and MD-11 aircraft. The Soviet international airlines understood what kind of equipment they needed to be competitive with foreign carriers flying into the Soviet Union. They also earned some hard currency from ticket sales abroad and, in the case of Aeroflot International, from selling jet fuel in Ireland at world prices. Leases would often be necessary to make the deal happen, and certainly sales financing would be very important. In both cases, there were significant risks of commercial default, adverse exchange rate fluctuations, and indeed political risk—as happened when the Soviet Union broke into its separate parts, when the Communists threw out Gorbachev in their *coup d'état*, only to have Boris Yeltsin overcome them in turn. Export financing agencies such as the Export–Import Bank and government export insurance offered by the Overseas Private Insurance Corporation would play important roles in such sales and leases.

A second strategy was to find a way of installing GE engines in Soviet airframes. There were possible applications for GE engines inside Russia for the CT7 turboshaft engine in new Mil and Kamov helicopters, and CT7 turboprop engines in a new small Sukhoi airliner. But larger Soviet airliners could also gain in performance and reliability from engines such as the CFM56 and the CF6. We assumed that Soviet-designed and -built aircraft would of necessity continue to satisfy much of the market. There would always be a preference for local products by local buyers, who could pay for them in rubles. Putting GE engines into such aircraft would command a given portion of the market without subtracting from engine sales GE could make in Boeing or Airbus Industrie airliners.

The Russian airframe companies recognized the advantage of a more efficient engine: an Ilyushin 86 airliner that, flying from Moscow to New York with Kuznetsov engines had to make a fuel stop in Ireland and sometimes in Newfoundland as well, could make the flight nonstop with CFM56 engines. On given routes, the CFM56 engine would save 20% or more of the fuel expense and comply with strict noise limits imposed at foreign airports by the International Civil Aviation Organization. On the debit side, Ilyushin would have to design a new pylon to mount the engine and would have to go through a flight test program and certification to international standards. Those nonrecurring costs plus the price for CFM56 engines—more expensive, of course, than the old Kuznetsov engines—would have to be recovered through savings in fuel consumption and other performance advantages. The Ilyushin design bureau was not enthusiastic. It had its eye on a newer airplane, the Ilyushin 96, and for this it wanted the Pratt & Whitney PW2037, the same model used on the Boeing 757. Ilyushin planned to lengthen the Il 96 and to equip it also with a Honeywell flight control system. The design bureau expected to have an aircraft whose performance would be competitive with that of Boeing and Airbus Industrie aircraft of the same size. Pratt & Whitney offered to make available 20 engines for the program as its share of the investment, their cost to be recouped as Ilyushin sold aircraft. One problem was that Ilyushin expected to export the aircraft, while Pratt & Whitney insisted that Aeroflot would have to buy some of them for use inside Russia to facilitate subsequent export sales. Export sales would of course compete with sales of Boeing and Airbus Industrie airliners. In any case, because money is short for aircraft development and the airlines have shown a distinct fondness for buying or leasing foreign equipment, there has not to this writing been much progress on the Il 96 and none on the Il 86. Nor has Rolls–Royce made much progress with its plan to install the RB.211-535 in the Tupolev 204, an aircraft similar in size to the Boeing 757.

The third branch of the GE strategy was to find a program of cooperation with the Russian engine industry. If that succeeded, it would make the Russians partners rather than competitors for the internal Russian market and participants in the market outside Russia. It was a strategy that had worked well for GE in Germany and France with the CF6, and had led to the CFM56 and GE90 cooperative programs later. Making this strategy succeed requires selecting the right partner and defining a program attractive to both sides. One of the companies GE looked at was Motorsich, a huge factory in Zaporozhye in Ukraine. Motorsich was associated with the Progress Design Bureau and made the D18 and D36 turbofan engines, and smaller engines for the Yakovlev Ya 42 commuter and executive aircraft. Motorsich was already working on a development of the D36, the D436, a high-bypass turbofan of about 14,000-lb takeoff thrust, roughly the size required by the 100-seat airliners under discussion in Europe and Asia. Motorsich would have been delighted to go into a cooperative program, and offered to have half of the engine manufactured by a European or American partner. The partner would also be in charge of marketing and product support, an area in which Motorsich was prepared to defer. Selling a Ukrainian engine to a Western airline would have required that a credible Western manufacturer stand behind the product and assume total responsibility for its economy and reliability.

SNECMA was completely against such a notion. It intended to develop its own engine in this size and did not want the D436 to become a competitor or distract from a new engine program. At that time, Ukraine became an independent nation, and GE realized that an association with a Ukrainian engine firm would not have had any beneficial impact on the bigger market in Russia. GE had much more to lose with SNECMA than it might have gained with Motorsich and dropped the discussions.

A potential partner should possess three desirable attributes: technical capability, clout in the market place, and access to funding for development programs. GE found a company with experience in large commercial engines in Perm, the manufacturer of the

PS90 engine used in the Ilyushin 96. Perm is an old manufacturing town on the western side of the Ural Mountains, the site of an important cannon foundry during the nineteenth century and the site also of a notorious prison camp, Gulag 35, which had been closed shortly before GE began discussions with the engine company.

Our GE/SNECMA team flew to Perm from Moscow in a Yak 42 corporate aircraft owned by the factory. It was not the same as a trip in an executive aircraft in the West. At 6:30 a.m. on a winter morning, the old Domodedovo airport in Moscow used by corporate flights as well as other internal airline flights was crowded with people sleeping, waiting to go somewhere. The toilet in the terminal building was easy to find: just follow your nose. Our Russian pilot scraped the ice from the windshield and fired up the engines, AI20s from Motorsich in Zaporozhye. Crates of parts and other supplies were piled loosely in the aisle next to the exit door. We took off into the dawn at 7:30 a.m. and began to climb to cruise altitude. As soon as the plane reached 10,000 ft, a steward brought plates of cold cuts and drinks. Beer. White wine. Vodka, of course, in several flavors. Russian brandy. By the time the aircraft landed in Perm, it was already dark. The head of the design bureau and the factory chief met us on the runway with cars to take us to a hotel in town.

Perm wanted help in modernizing the design of its PS90 engine and in raising its power. GE and SNECMA engineers participated in several design reviews with Perm, and some good ideas were exchanged, but little of substance was accomplished. Both GE and SNECMA made clear that they were not going to give money to Perm to develop a new engine. GE took a brief look at the notion of licensing Perm to build the CF6-50 engine for Russian applications and the possibility of having engine parts made in Perm for GE's production programs, provided that the price would really be as low as the Russians kept saying. These were ideas similar to those GE had pursued with some success in China 10 years earlier, but they were too ambitious for those times. The world airline market had been in a deep recession ever since the Gulf War of 1990; sales were down from their 1991 peak, and GE Aircraft Engines needed to reduce its size and overhead expense to match the expected reduction in revenue. There was an instant excess of manufacturing capacity in the United States and Western Europe, and no one wanted to spend the effort and money required to set up a new parts supplier, let alone one with whom a complete new understanding of specifications, technical data, and quality systems had to be developed. Nothing came of the ideas.

Pratt & Whitney held similar discussions with Perm in conjunction with the Ilyushin 96 program, with a similar lack of results. The Russians were bitterly disappointed: neither of the American companies had brought them any tangible programs or work.

GE's chairman, John F. Welch, Jr., visited Moscow in 1991 to explore opportunities not solely for aircraft engines but for all GE businesses. Markets were growing for plastics, light bulbs, and medical x-ray equipment through Russian distributors. Power generation and the oil and gas pipelines were a traditional market: GE had worked on hydroelectric turbines at the Dnepropetrovsk dam as far back as 1936. There was strong competition from Siemens and ABB, powerful European firms with wide experience in the former USSR. Welch had reservations about the economic viability of installing the CFM56 engine in the Ilyushin 86 and about partnership with Russian engine companies, but, in a meeting with Gorbachev and Belyakov, his minister for "conversion" of the aerospace industry from military to civilian production, Welch promised to help train Russian managers in the skills they would need in a new demand economy. GE Aircraft Engines selected some three dozen Russian engineers and managers from the airlines, aircraft design bureaus and factories, and from the engine companies, to attend a special three-week training program at GE's management training institute in Crotonville, New York, followed by a visit to the GE engine factory in Evendale, Ohio. There were instructors from GE and several business schools and guest speakers from the U.S. Commerce Department and the FAA. One of the highlights of the Crotonville course came when the Russian group was taken into an auditorium filled with other GE people undergoing management training in Crotonville at the same time. Jack Welch stood in front of the auditorium in his shirt sleeves and took questions from the audience, mincing few words in his answers. The Russians understood perhaps 1% of what was said, but they understood 100% of the body language: "We need a chairman like that!" The Crotonville course cost GE almost $400,000.

The outcome of all of these initiatives has been that after Aeroflot split into several airlines, these have bought Boeing and Airbus Industrie airliners, some of which come with GE and CFMI engines, supported by GE and CFMI technical representatives. Many of the service representatives are Russian engineers trained in the United States and France. GE has begun to place a few hesitant parts orders with Russian manufacturers. The cost of establishing a common technical language is as high as everyone predicted, and the price savings not as big as GE would like. There is also a modest joint venture with an engine company in Rybinsk, about 200 miles north of Moscow, covering the CT7 turboprop engine for small airliners and the LM2500 engine for electrical power generation. This is more likely to succeed than some of the grandiose predecessors. Ilyushin has not gone ahead with installation of the CFM56 in the Il 86. The Il 96 program with the Pratt & Whitney PW2037 has gone ahead fitfully, with talk of as many as 20 aircraft for Aeroflot—but there is as yet no sign of more than prototypes. Market development in China took 15 years to bear fruit in major sales; it will take at least as long in Russia, whose economy will also need that time to stabilize.

In fact, there is little anyone can teach the Russians about the technology or manufacture of aircraft engines. Their own capability in this field is significantly stronger than is the case in China or Japan, the results of decades of investment in research and development. The Russians know as much about aerodynamics and metallurgy as anyone and have manufactured their engines in large numbers. In the past, Russian designers were forced to design with big safety factors to compensate for the large variability of physical properties of their materials. Given access to materials with more consistent physical properties, they will be able to design lighter and more efficient parts with the same reliability as before. Their large-scale test facilities and wind tunnels are unique (Boeing set up an aerodynamic design office in Moscow to take advantage of some of the Russian brainpower available). The tricks of how to design for maintenance, and how to take advantage of materials with more consistent physical properties than those they have had in the past, are mastered soon enough. What the Russians have lacked is a thorough understanding of cost and price, and particularly of how to market a product and then support it in service so that the customer is encouraged to come back for another. When they have mastered the economics, the West will lose its unique comparative advantage. When the Japanese set out to learn about turbojet engine manufacturing after World War II, they adopted the humble attitude of an apprentice. The Russians were among the victors of the war and a superpower thereafter. They believed, sincerely or not, that they did not need to feel humble about their capabilities. What would make them competitors at the world level would be an alliance between equals: Russian technology partnered with the manufacturing skills, marketing, and product support expertise of . . . whom?

Of all potential partners, the Japanese may have the most to gain from a partnership with the Russians and the least to contribute. They are also notoriously averse to risk. The American engine industry has the most to contribute but also the most to lose from real competition by the Russians. American and European companies have tried several strategies of international collaboration. For a long time, Boeing has done its best to protect its market position by keeping close control over its own expertise. Boeing has brought in other firms, notably in Japan, that may become competitors in the future, as subcontractors responsible for their own technology. Boeing's motives were access to external program funding for the parts assigned to the subcontractors, and preferential market access in their countries.

GE has in the past been more open to international cooperation than have Pratt & Whitney or Rolls–Royce, on the theory that GE had more to gain in market share and from market expansion than it might lose from a transfer of expertise to a potential competitor. That strategy worked well for GE, and the success was noted and copied to some extent by Pratt & Whitney and Rolls–Royce, and recently even by SNECMA and MTU.

If history is any guide, GE would be the most likely to enter into a cooperative agreement with a Russian engine company, along the lines of CFMI. GE's objectives would be to gain a dominant share of the Russian market; to make use of special Russian technology, particularly in compressor aerodynamic design and major Russian test facilities; and, as long as wage differentials exist between Russia and the West (perhaps for the next 5 or 10 years), to gain the benefit of low-cost Russian engineering and manufacture. The critical determinant will be the availability of Russian funding for the Russian work share; a tough hurdle given the lack of liquidity in the Russian economy.

Chapter 18

Product Planning and Partnership— Mixing Oil and Water

THE economic life of a successful engine design is perhaps 30–40 years. Advanced product planning involves thinking about the technology required for a superior product far in the future and putting people and resources in place to create the technology, sometimes along more than one path.

In the 1980s, the airlines began to study big twin-engine aircraft, with passenger capacity greater than that of the DC-10 and L-1011 trijets and with good long-range performance. The objective was a seat–mile operating cost lower than that of the 747 as well as lower aircraft–mile operating costs. The new twins would have a two-man flight deck, made possible by "fly-by-wire" electronic controls that allowed the airline to eliminate the flight engineer, the third person in the older cockpit. Two engines are less expensive to buy and operate than three, let alone four. The new turbofan engines had proved themselves so reliable that the airlines and the national airworthiness authorities had little concern about long transoceanic flights by twin-engine aircraft. And the airline routes with the greatest growth rate were along the east coast of Asia, generally within easy reach of airports for diversion in an emergency.

Airbus Industrie was the first off the mark. It believed that Boeing, having no direct competition for the 747, could charge high prices for the type and use its profits to cut prices on the 767, which competed directly with the Airbus Industrie A310. To cut into the quasi-monopoly of the 747, Airbus Industrie launched a big long-range twin, the A330. This it offered with any of three engines: the GE CF6- 80E, the Pratt & Whitney PW4000, and the Rolls–Royce Trent, a growth derivative of the RB.211-535. To press the 747 even further, Airbus Industrie made imaginative use of the same fuselage and wing to offer a four-engine aircraft of very long range, the A340, with four CFM56-7 engines.

Boeing countered the A330 with an even bigger twin, the 777, for which it completely overhauled its design control and manufacturing system. It is the extreme automation of the manufacturing process for the 777 that Boeing expects will protect its lead in this market niche. To a considerable extent, specifications and designs for the 777 are in electronic form in a computer, rather than on paper. Identical data are instantly available to the design engineer, the tool designer, the production crew, and the quality control team. Boeing's subcontractors and partners have immediate electronic access to the same data. Designers can test the fit and interference of structural parts, tubing, and accessories on their computers before the first piece of metal is cut. Boeing built a vast new factory to assemble the 777, to which parts come from suppliers in the United States and Japan. When aircraft fuselage parts are made in the conventional way, separate dies and inspection fixtures, very expensive because of the precision required, are needed for each part. Setting them into presses for sheet forming and into other machines for automatic drilling and riveting takes time and costs money; storing those not in immediate use requires a large amount of warehouse space. Some of the Japanese manufacturers, who learned how to work with Boeing on the 767, have developed computer-controlled tooling that can be adjusted immediately from one part number to the next without the need for rigid tooling for each part number or warehouse space to store it. The savings in tooling costs and warehouse space are immediate and great. Equally important, a press or drilling and riveting machine can be set up in minutes by the computer for a different part number, allowing the manufacturer infinite schedule flexibility and the ability to make production batches of a single part, reducing the inventory of work in process and, therefore, the working capital tied up. Boeing plans to employ this system for other models as well, particularly for the high-production

737, and expects to be able to reduce the manufacturing cycle time between receipt of a purchase order to delivery to the airline to less than one year—it took as long as three years in the 1980s.

For the Boeing 777, Pratt & Whitney offers the PW4084, Rolls–Royce the Trent 800, both growth versions of existing engines. But GE has taken a daring step and has designed a completely new engine. The GE90 (Fig. 10) uses an advanced compressor from the E Cubed technology program and a 123-in. diameter slow-speed fan with composite blades and a bypass ratio of 9 or 10:1, significantly higher than the bypass ratio of its competitors. The fan and its cowling have roughly the same diameter as the entire fuselage of a 737; such a big engine is not easy to mount under the wing of an airliner, whose landing gear must be long enough to allow the lower lip of the engine cowling to be clear of the runway surface. All three engines are in service, certified at about 85,000 lb of takeoff thrust. After the sale of 335 aircraft, the Pratt & Whitney engine has some

Fig. 10 GE90 in a British Airways Boeing 777. (GE negative)

39% of the orders, Rolls–Royce 33%, and GE about 28%. Each claims growth potential to a thrust rating of more than 100,000 lb, ratings probably required for growth versions of the 777.

But the big twins are only one part of the large airliner market. As the world economy recovered in 1995–1996, airline passenger traffic began to grow again and the aircraft product planners began to talk about even larger aircraft. Airbus Industrie proposes the A3XX, a 500–600 passenger aircraft to be certified in 2003; Boeing talks about an aircraft with a takeoff weight of 1 million lb and a 600-passenger capacity, the 747-600. But, in the meantime, Boeing is studying an intermediate growth of the 747, the 747-500, with 450–500 seats, a new wing, and perhaps the 777 fly-by-wire cockpit. Such an aircraft could be certified in 2000, ahead of the Airbus Industrie A3XX. Boeing believes that the larger part of the market for such very big aircraft will be at the 450–500 seat end of the range and can be served by the 747-500, the growth model of the current 747, more quickly and for less investment than by the A3XX. Because it has no current equivalent to the 747, Airbus Industrie is forced to start with a clean sheet of paper and, therefore, might as well design an airliner larger and more advanced than the 747-500. That will cost more and take longer, but Airbus Industrie believes that there will be a substantial market at the upper end of the range, and that the A3XX can capture it before Boeing is able to develop a 747-600.

Each of the airframe companies claims that the market forecast justifies the strategy it has chosen to follow. Each has been successful in the past with such a product planning process, Boeing by extracting the maximum revenue from its existing product line before it offers a newer and more advanced aircraft, Airbus Industrie by leapfrogging existing Boeing models with completely new advanced performance designs. That is the strategy to follow to overtake the market leader, and it requires patient investment capital and the tenacity to stay the course.

The challenge to the engine companies is to accommodate such growth at minimum technical risk and expense. All three have offered engines at thrust ratings of 85,000 lb for the 777 and A330, with the potential to grow to 100,000 lb of takeoff thrust. But the growth of the 747 and A3XX will require engines of an entirely different size. These are both four-engine airliners and they will need engines rated at 70,000–80,000 lb of thrust—smaller than the new engines for the 777 and A330 but with specific fuel consumption and low exhaust emissions at least as good. Airbus Industrie stated that it had an agreement with GE for such an engine at 75,000 lb, scaled down from the GE90, but that seems to have come to a halt. Then Airbus Industrie announced that Rolls–Royce agreed to develop a derivative of the Trent. Because the PW4084 and Trent 800 were originally scaled *up* from smaller versions, creating 75,000-lb thrust versions of advanced performance appears to be a relatively simple task. Scaling *down* the GE90 may be more complicated. It would need a new fan, and the associated changes in the low- and high-pressure compressors and turbines. And the fundamental question is always whether the market will be big enough to justify the cost of development.

All of this has led to an astonishing jockeying in the engine industry. Citing the need to limit investment in face of a limited market, GE and Pratt & Whitney set aside their rivalry and announced a joint venture to develop a 75,000-lb thrust engine, the GP7000, building on the technology of the GE90 core engine and that of Pratt & Whitney's fan. GE forecasts a market for airliners with 500 or more seats of 535 between the years 2000 and 2014, Rolls–Royce 500, and Boeing 470. Airbus Industrie, on the other hand, sees a potential market of 1380 airliners of this size. If the engine company forecasts are correct, there will be a total market of about 2000 installed engines and perhaps 400 spare engines—enough for one engine model to be financially successful, but not two, and certainly not three. Therefore, the GE/Pratt & Whitney joint venture may have as one important objective reducing the number of engines competing for this limited market from three to two. A reduced investment by each of the GP7000 partners (a mere $1 billion each, according to published reports) could lead to a profitable business plan, while to allay any concerns about a monopoly position, there would always be the new Rolls–Royce Trent 900. In November 1996, Sir Ralph Robins, the chairman of Rolls–Royce, in response to the GE/Pratt & Whitney collaboration, visited Paris to offer SNECMA a partnership share in the Trent 900.

One may be permitted a degree of skepticism about all of these announcements of collaboration and partnership. They have their rationale, to be sure. Logically, product-specific cooperation between competitors ought to be possible when the common objective is so compelling that it overcomes ingrained habits of competition between other engines in the product line. But much of the experience of the last 20 years suggests the opposite. Pratt & Whitney and Rolls–Royce were unable to create a JT10D as a competitor to the CFM56 because they competed on every other program; GE and Rolls–Royce could not collaborate on the RB.211-535 and the CF6-80C because Rolls–Royce insisted on its independent development of the Trent to compete with the CF6-80C. Pratt & Whitney did manage to collaborate successfully with Rolls–Royce on the V.2500; Rolls–Royce was forced into the collaboration by Japanese pressure, and Pratt & Whitney had nothing to lose and something to gain against the CFM56.

A partnership between Rolls–Royce and SNECMA could pose some risk to old and successful alliances, such as that between GE and SNECMA on the CFM56. It is not easy to reconcile a SNECMA/Rolls–Royce alliance with SNECMA's participation in important GE programs such as the CF6-80 and the GE90. The stream of revenue and income represented by the CFM56 will continue but will also continue to require further research and development investment. There will be a need for an engine in the 35,000 to 45,000-lb thrust class, for a medium-sized twin-engine aircraft following the Boeing 757 and Airbus Industrie A321 and A310, an engine that could also be used for a growth version of the A340. As a rough rule of thumb, to justify its development, such a new engine should be at least 15% better in performance than the ones it replaces. For SNECMA, working on such an engine with GE while working with Rolls–Royce on another engine that competes for resources would be a delicate and expensive task. From that point of view, the GP7000 has a Machiavellian benefit for Pratt & Whitney: it drives a wedge between GE and SNECMA and reduces the competition to Pratt & Whitney from new versions of the CFM56.

The history of cooperation between the big aircraft engine companies contains as many examples of failure as of success. GE enjoyed success in its CF6-50 co-production program with SNECMA and MTU and later with Volvo and FIAT, and the SNECMA/GE joint development of the CFM56 led to one of the world's most successful programs of co-production and joint marketing. But other GE co-operative ventures were not at all successful.

In the early 1980s, after GE had withdrawn its CF6-32 from the Boeing 757 market, it began discussions with Rolls–Royce about an ambitious plan for cooperation. GE would take a 15% share in Rolls–Royce's RB.211-535 program, an engine size in which GE no longer had an equivalent model of its own, and Rolls–Royce would take an equivalent share in the CF6-80C program, an engine larger than the RB.211-22 which Rolls–Royce developed for the Lockheed L-1011 Tristar. Each company had the option to enlarge its share to 25%, and neither company would have to develop an equivalent engine on its own. GE and Rolls–Royce signed agreements with the explicit assent of the U.S. Justice Department, and the engineers and manufacturing specialists of the two companies began to work together with enthusiasm and quick results.

There are three major driving forces for such cooperation: access to development funding; access to proprietary technology, or at least elimination of the need to duplicate it; and access to protected markets. Given Rolls–Royce's and GE's proven capability, the access each had to funding for its own work, and the strong position each held in certain geographic markets, the advantages of their working together on two engine models that complemented each other's product range were clear. The concept appealed to Sir Douglas Duncan, Rolls–Royce's chairman, who had grave reservations about the economic wisdom of investing alone in the development of a new engine larger than the RB.211-22. Duncan was not an engineer and had been appointed to his post by the British government, Rolls–Royce's owner since the 1970 receivership. His role was to ensure the continuing commercial existence and success of the company, and he had no personal ego invested in Rolls–Royce's primacy as a designer of world-class aircraft engines.

Others in Rolls–Royce felt quite differently. Like Pratt & Whitney and GE Aircraft Engines, Rolls–Royce is managed predominantly by engineers who have come up through its ranks. The company has an illustrious history of technical innovation and leadership in the field of aircraft engines beginning in 1915, in which many of the managers played a part. They resented yielding any part of the commercial engine market to GE, a firm that had not begun to develop aircraft engines until 1942. They were convinced that Rolls–Royce could develop a big engine for the 747 better than GE's CF6-80C and do it for much less money than Sir Douglas Duncan assumed in his strategic calculus.

Opinions were mixed also in the British government, where some saw Rolls–Royce as a shining example of high-technology industry, able to command a significant share of the world market, and worth government investment both for that role and as an example to other British firms. The technical experts in the British Ministry of Defence, who funded the development of Rolls–Royce's advanced technology and who had to pay monopoly prices for the production of military engines and spare parts for the RAF, were less sure than the politicians about Rolls–Royce's ability to make a commercial success of a new big engine for the 747, given the competition from Pratt & Whitney and GE.

Suddenly Sir Douglas Duncan died of a heart attack and was replaced by Lord Tombs, whom the government charged with improving Rolls–Royce's balance sheet so that the company could be privatized. Under Tombs, Rolls–Royce began the development of its own new large engine, the RB.211-524. GE protested that this contravened the spirit of the agreement to cooperate on the CF6-80C. While the agreements did not forbid Rolls–Royce to develop the new engine, developing an engine to compete directly with the CF6-80C created a conflict of interest: as a partner of GE, Rolls–Royce would be privy to marketing information and price concessions for the CF6-80C, information that was bound to affect its own marketing of the RB.211-524 to the same customers. The arrangement would be likely to violate American antitrust laws and would certainly violate the spirit of cooperation on the CF6-80C.

Rolls–Royce and GE parted company, to the dismay of the engineers and manufacturing people in both companies, who had found it very easy to work together. Some money changed hands between the companies in compensation, and it took several years to disentangle the long-range plans that had been made for production of engine parts. After Rolls–Royce was successfully privatized, Tombs was succeeded by Ralph Robins, who had been a strong advocate of the RB.211-524.

There was a parallel case in Germany. There MTU produced J79 and T64 military engines for the German government under license from GE and had taken a 16% production share of the CF6-50 during the launch of the Airbus Industrie A300. These programs provided a significant workload for MTU's factories and had been financially profitable, but they held comparatively little appeal for MTU's technical staff, the German engineers who were the inheritors of a proud engineering tradition that carried back to the BMW003 turbojet engine for the Me 262 fighter in World War II. Sure that they were capable of playing a leading role in the design of new aircraft engines, they chafed at their diminished responsibility in programs limited to cooperation in production. The German government, emphasizing *European* cooperation as a critical political factor in German NATO participation, encouraged MTU and other German companies to work with Rolls–Royce rather than the Americans on military engine technology. Engineers at MTU developed close working relationships with Rolls–Royce, a collaboration that produced the RB.199 engine for the Tornado and the VJ200 engine for the Eurofighter. Pratt & Whitney took note of the important role MTU had played in getting the German government's assent to the use of the GE CF6-50 as the launch engine for Airbus Industrie and began to cultivate a relationship with MTU by placing production orders for JT8D spare parts when the workload of parts for J79 and T64 engines began to run down.

When SNECMA and GE launched the CFM56 as the successor to the JT8D, MTU wanted to join in an engineering role but was rebuffed by SNECMA, which wanted to solidify its own primacy in Europe immediately below the first rank of aircraft engine makers. Pratt & Whitney immediately offered MTU a share of the development of the PW2037, making it responsible for the design and production of the entire low-pressure turbine. Later, MTU also joined the V2500 program when it was launched as a competitor to the CFM56 by Rolls–Royce, the Japanese, and Pratt & Whitney.

266 Product Planning and Partnership—Mixing Oil and Water

As had been true at Rolls-Royce, there were two schools of thought in MTU. The production and financial people, with the experience of several profitable programs led by GE, favored continuing such cooperation and caused MTU to join the CF60-80 co-production program. But MTU's design engineers had worked more with Rolls-Royce and Pratt & Whitney than with GE and had ambitions that went beyond specializing in the design of low-pressure turbines. GE recognized this and, when it launched the new GE90 engine for the Boeing 777, invited MTU to join the program in an engineering role as well as in production. GE extended the same invitation to SNECMA, FIAT, and IHI. There was an obvious advantage to having such potent partners in key European and Japanese markets, besides leaving fewer potential partners available to Pratt & Whitney or Rolls-Royce. The novelty of the engine technology and the expanded role appealed to MTU, and it signed a revenue-sharing agreement with GE, as did SNECMA, IHI, and ultimately FIAT.

In the consolidation of the German defense industry at the end of the 1960s, MTU had become a division of Deutsche Aerospace, itself a part of Daimler-Benz. In the early 1980s, the chairman of Daimler-Benz, Edzard Reuter, developed a corporate strategic vision of a global alliance that would encompass the car and truck business as well as aerospace markets and reached a broad agreement with United Technologies Corporation, the American parent company of Pratt & Whitney, and with Mitsubishi Heavy Industries, Ltd. Reutter talked about cross-shareholding between Pratt & Whitney and MTU and their collaboration in a broad range of aircraft engine programs, including the PW4000 for the Boeing 777. GE protested that the PW4000 deal was a direct breach of its own agreement to cooperate with MTU on the development and production of the GE90. The supervisory board at MTU approved the GE90 agreement, but Daimler-Benz claimed that its own board had not been asked to approve and that it had the power to nullify the MTU agreement. Reuter made light of the matter and accused GE's chairman Welch of exaggerating its seriousness. GE promptly filed suit against MTU for breach of contract in U.S. Federal Court in New York. This is a course of action less common in Europe than in the United States, and when it became clear to Daimler-Benz that GE was serious, the suit was settled before it came to trial. MTU returned all the GE90 data to GE and promised to forget all about them in any dealings MTU had with Pratt & Whitney. In addition, Daimler–Benz paid damages to GE in excess of $100 million.

The lesson to be drawn from such examples is that collaboration on multimillion dollar programs is inherently difficult. Success demands the whole-hearted and unreserved support of each partner's top management and the enthusiastic cooperation of their organizations. If there is a fundamental conflict of interest between partners, particularly if any of them has its own product competing with the product of the collaboration, collaboration is bound to fail.

CHAPTER **19**

■

The New German Engine

THE market for large airliners, we have seen, is dominated by Boeing and Airbus Industrie, with a minor market share falling to Douglas for a time until it was absorbed by Boeing. Not since Airbus Industrie was founded has anyone been willing to compete directly against the leaders in the market segment on which they have concentrated—airliners with 150 seats or more. Companies with ambitions to build airliners, and governments willing to sponsor such ambitions, have looked for market segments that the giants have neglected. Thus, there has been much talk in Europe about a 100-passenger airliner, requiring two engines each with 14,000 lb of takeoff thrust, with growth potential to 18,000 lb, but the talk has not been followed by action, reflecting not only the slow state of the market but great uncertainty about the aircraft itself. The Dutch aircraft company Fokker, owned jointly by the Dutch government and Deutsche Aerospace, was in this market with the Fokker 100, powered by the Rolls–Royce Tay. The Tay uses the core of the old Spey engine with a new high-bypass fan. It is not a modern engine but, like the Pratt & Whitney JT8D in its day, it has been alone in its size range. Fokker, however, went bankrupt because it could not sell enough airliners to cover its costs and because, in the absence of a credible market, Deutsche Aerospace refused to provide more working capital. The Germans have evidently given up on their ambitious plans of five years ago for a family of new 80–120 seat airliners.

Other external investors, such as Korea's Samsung Aerospace, have also decided that Fokker does not represent a sound investment. Although it has been rumored that the Russian airframe firm Yakovlev has expressed interest in buying Fokker, Fokker's market value is nil if no credible buyer will pay anything for its assets, organization, and goodwill. Yakovlev could benefit from the use of Fokker's marketing and product support organizations to sell its own aircraft but probably cannot provide the hard currency required for an operating budget.

This market segment can obviously be served by the MD-80, the Boeing 737, or the Airbus Industrie A320 at less than full passenger load, at some sacrifice in airplane–mile cost but with relatively low production costs that allow a sharply competitive selling price and provide commonality with existing fleets. Airbus Industrie is also offering a shortened version of the A320, the A319; Boeing is offering a whole range of new 737 models with a new wing, the -600/-700/-800.

Douglas waved its magic wand once more over the corpse of its existing design and shortened the MD-80 fuselage to create the MD-95. Because the MD-95 is smaller than any aircraft in Boeing's product range, it was the only Douglas model other than the MD-11 to survive the merger with Boeing and has been rechristened the Boeing 717. However, it needs a truly new engine in place of the old JT8D. Douglas received a launch order for 20, with options for 20 more, at a price so low that Airbus Industrie and CFMI, which had proposed the A319, backed away during the final price auction. The order was from Valujet, a new low-cost airline itself in some trouble. It lost an aircraft in a tragic accident at Miami in 1996, blamed in part on ValuJet's lax supervision of a maintenance subcontractor. Since then the airline has had trouble attracting passengers and, in late 1997, changed its name to that of another small airline it was acquiring so that it could sell tickets without frightening passengers away. A tough-minded investor would worry about a launch order for a new aircraft model from an airline in that position. For the moment, Boeing is keeping the 717 on the market.

In the 1970s, Japanese planners conceived the RJ500 engine. They saw a market opportunity for a high-bypass turbofan smaller than the CFM56 and more modern than the JT8D and the Rolls–Royce Spey and Tay. Until then, the Japanese played a minor role in the market for commercial aircraft engines. They were looking for a niche in the market not yet contested by the "Big Three," Pratt & Whitney, GE, and Rolls–Royce. As time went on, the RJ500 lost the Boeing 737 application to the CFM56 and evolved into the V2500, a larger engine that has since become a direct competitor to the CFM56.

Some 15 years later, this sequence of events is being played out again in the market niche just below the CFM56 and V2500. The market for engines of 14,000–16,000 lb of takeoff thrust has been dominated by the Rolls–Royce Spey and its refanned version, the Tay. The Ukrainian engine firm Progress has offered a modern engine in this size, the three-spool D436, but the conservatism of the world's airline industry inevitably rejects an engine from a new supplier with whose products only Aeroflot has had experience. Neither GE nor Pratt & Whitney has so far offered a completely new design in this size, perhaps because the two engine companies doubt whether the market is big enough to justify the development cost or perhaps because their priorities are with the larger engines. Whatever their reasons, as long as this is the case, Rolls–Royce has the market segment to itself and no incentive to replace the Tay with a new product. The largest applications for the Tay have been in Fokker 100 airliners and Gulfstream business jets. Deutsche Aerospace has always been interested in the 100-seat airliner market as a segment not yet dominated by the Americans or the French and bought control of Fokker both to enter the market and reduce the number of competitors for an airliner of that size, a move that was unsuccessful, as we have seen.

The Douglas MD-95/Boeing 717 is still on the market and clearly needs a new high-bypass engine, more modern than the old JT8D or the Tay. Happily, such a new engine has appeared: the BMW/RR 715, under development by a joint venture between BMW and Rolls–Royce, who have offered the same engine for the Gulfstream and the long-range version of the Canadair Challenger.

Bayerische Motoren-Werke is an old name but a newcomer on the commercial aircraft engine scene. During the 1930s, BMW built

Pratt & Whitney engines under license and, during World War II developed its own large engines, including the BMW801 for the Focke–Wulf 190 fighter and the BMW003 turbojet engine for the Me 262 jet fighter. In 1959, owned by the BMW automobile company, BMW Triebwerkbau GmbH began to manufacture GE J79 engines for the German air force's F-104 program. The parent company, intent on its core business and strapped for working capital, sold the aircraft engine division to MAN Turbo (in 1960, the automobile company offered the engine division to GE, contingent on a $4 million loan and equity investment in the auto business; GE turned the offer down), which sold it to Daimler–Benz, which merged it with its diesel engine division as Motoren- und Turbinen-Union, or MTU. MTU is the major contractor for the postwar German air force's military jet engine programs, building the GE J79 and T64, collaborating with Rolls–Royce on the development and production of the RB.199 and VJ.200 jet fighter engines, and is also a production partner for GE on the CF6-50 for the Airbus Industrie A300 and the CF6-80. Before Daimler–Benz bought the company, its managers tried to participate in the CFM56 program and discussed the possibility with SNECMA in Paris. Under its agreement with GE, SNECMA had the deciding voice over European partners, and seeing no advantage in admitting the Germans treated them rather coldly. The German managing director and his projects manager returned to Munich furious at having been snubbed.

When Pratt & Whitney invited MTU to participate in the development of the PW2037, the Germans were quick to agree. Pratt & Whitney offered MTU a chance to participate in design and development, while its role in GE programs was limited to production, which might be lucrative but did not do much for the ego of German design engineers. MTU went with Pratt & Whitney and designed the low-pressure turbine of the PW2037. When Pratt & Whitney then joined Rolls–Royce and three Japanese engine companies in the V2500 program to compete with the CFM56, they all invited MTU and FIAT to join International Aero-Engines, Ltd. In revenge for the rejection by SNECMA a few years before, MTU accepted in revenge for its rejection by SNECMA a few years before. Also joining was FIAT, although later it tried unsuccessfully to edge out when sales were very slow and FIAT's parent company had come under severe financial pressure from the lagging auto business.

Meanwhile, BMW, having earlier sold off the aircraft engine division that became MTU, had become very successful and profitable in its automobile business and wanted to reenter the aircraft engine field to diversify against slow growth on the automobile side and infuse aerospace technology to its cars. Similar notions impelled Daimler–Benz in its mergers with various German aerospace companies and in the grandiose alliance with Mitsubishi and United Technologies Corporation, the parent company of Pratt & Whitney (Richard Ridge of GE has noted the irony that General Motors disposed of its aircraft engine company, Allison, at about the same time; perhaps automotive companies diversify during economic booms and divest during busts, responding to cyclical external economic conditions rather than long-term strategies). BMW retained the services of the American management consultants Booz, Allen & Hamilton to advise it on entering the aircraft engine business. Booz, Allen told BMW that it should take an experienced partner and listed eight candidates. GE's name was at the head of the list, Rolls–Royce's second. Thereupon BMW asked GE whether it had any interest in joint development of a new engine in a thrust range where there was then no real competition, around 14,000 lb.

There are two applications for such an engine. The first is a 100-seat airliner of the type considered by Fokker, Deutsche Aerospace, Bombardier, the Japanese, and the Chinese. The second is the perennial European military transport to replace the Transall and Lockheed C-130 in the air forces of Germany, France, Britain, and Italy. Their combined requirement is perhaps 300 such future large airlifters (FLA). Perhaps another 200 FLA might be sold to other air forces. The FLA program would require 2000 installed engines, another 500 as spares, and would qualify for government development funding. Such a military engine program ought to be hugely profitable. BMW saw it as a great market opportunity for a European company that alone would be eligible for European government research and development funding.

The new aircraft engine division of BMW had been created out of whole cloth. A management and technical team was recruited from the engineering faculty of the Technische Hochschule-München and, for its production the company bought the gas turbine factory of Klockner–Humboldt–Deutz (KHD), the only German military gas turbine enterprise other than MTU. KHD built the T53 helicopter engine under Lycoming license, developed a gas turbine auxiliary power unit for the Tornado, and collaborated with SNECMA on the design and manufacture of the Larzac turbofan for the Alphajet trainer, but was now running out of work. BMW proposed to divide the development and production work of the new 14,000-lb thrust engine evenly with GE and set aside land and a budget of $400 million to launch the program. Because this took place at the time of the unification of East Germany with the Federal Republic of Germany, there may have been tacit political support from the government for a new engine factory at Dahlewitz, near Berlin in what had been East Germany—the Bonn government was encouraging West German firms to take over former East German enterprises or to set up operations in the East. In the meantime, BMW need production work from whomever it selected to be its partner to load the factory, until the new program generated enough work of its own.

An engine like the one proposed by BMW comes close to the size of the CFM56, and GE felt bound to discuss the concept with SNECMA, whose partnership in their joint programs was obviously more important to GE than a new smaller engine. SNECMA reacted with some heat. GE had been the leader in all their joint programs for big engines. For smaller engines such as this one, SNECMA intended to play the leadership role and in fact had a study under way for an FLA engine, using the high-pressure core of a new fighter engine, the M88. SNECMA saw no good reason to include BMW in such a program, but, if there were compelling marketing reasons to add a German partner, BMW's role would necessarily have to be smaller than SNECMA's. That ended the discussion because BMW thought, not unreasonably, that its $400 million investment and lock on the German military market entitled it to a voice in the program equal to that of any other partner—in any case, not to a role smaller than SNECMA's. The BMW representatives closed their briefcases and departed, perhaps leaving GE to regret the disappearance of $400 million ready to be spent on engine development.

BMW went to the second engine company on the consultant's list, Rolls–Royce, and received a much warmer reception. Rolls–Royce had every incentive to protect its market position for engines in this thrust range and welcomed a partner supplied with money to develop a modern successor to the Tay. The two quickly formed a new company, BMW/Rolls–Royce, with headquarters in Berlin, to design and produce a new engine. In the meantime, BMW could make parts for the Rolls–Royce Tay. The old KHD factory was already short of capacity so that some of the production work in BMW's share remains with Rolls–Royce. Ironically, MTU has all the production capacity BMW needs, plus orders from the German government for military engine spare parts and contracts for the development of the engine for the Eurofighter. That could be very attractive for BMW. Daimler–Benz, having already cast Fokker adrift and regarding MTU as a sink-hole for money rather than a valuable resource, may be happy to pass MTU over to a more enthusiastic owner. BMW is in fact trying to take over MTU at this writing, but the two companies appear to be far apart on price.

Douglas picked the new BMW/RR700 for the MD-95. Gulfstream, which uses the Rolls-Royce Spey in its Gulfstream II and III executive aircraft and the Tay in the Gulfstream IV, plans to use the BMW/RR700 in the Gulfstream V; the Canadian company Bombardier, which makes the CL601 executive aircraft and the 50-passenger RJ airliner, both with GE CF34 engines, has chosen the BMW/RR700 for a new executive version of the CL601 capable of trans-Pacific range, although, for its proposed new 70-passenger airliner, Bombardier will use either the growth version of GE's CF34 or a new engine proposed by Canadian Pratt & Whitney and SNECMA.

BMW has proposed a turboprop version of the BMW/RR700 for FLA but, like the MD-95, the new military airlifter program is hardly on a firm footing and in fact has not yet been authorized. France has withdrawn from the FLA project. Money is tight everywhere and there are other options. The cheapest replacement for a Transall or Lockheed C-130 tactical transport remains a new Lockheed C-130, with Allison turboprop engines.

Although large questions about the market remain unanswered, the BMW/RR700 joint venture has gambled on the opportunity to take over this market segment from the Rolls–Royce Tay. Rolls–Royce has taken a less risky course, continuing to serve this market segment with the Tay, and, in 1995, Rolls–Royce also bought Allison. This former division of General Motors built the Spey under Rolls–Royce license as the TF41 for the U.S. Navy, is the producer of the turboprop engine for the C-130, and has certified a turbofan engine derived from that turboprop for the Brazilian EMB-145 50-seat airliner. Through Allison, Rolls–Royce can participate in sales of more C-130 aircraft and can offer derivative engines for the smaller airliners.

SNECMA and GE also continue to compete in this market and have proposed a lightweight version of the CFM56 for airliners smaller than the A319. GE continues to work on a growth version of the CF34, with a new fan, core compressor, and low-pressure turbine, aimed at a 70-passenger version of the Bombardier RJ regional airliner. Two Japanese companies, IHI and KHI, are participating in the development of this CF34-8. Pratt & Whitney Canada is also studying a new engine of this size and has invited SNECMA and MTU to join it in such a development.

Only BMW has placed its bets firmly on the new BMW/Rolls–Royce engine. Discussions continue between BMW and Daimler–Benz about BMW's taking over MTU. If they can agree on price, BMW would gain MTU's manufacturing capacity, military production for the Tornado, and military development for the Eurofighter. In that case, BMW would also take over MTU's commercial production for GE and Pratt & Whitney, as long as the two American companies do not launch new engines of their own to compete with the BMW/RR700. That could bring about tensions similar to those that arose when Daimler–Benz directed MTU to join Pratt & Whitney in the PW4000 program after it had already joined the GE90. For the foreseeable future, the success of the BMW/RR 700 rests on the ValuJet order for MD95, Gulfstream, and Bombardier's Pacific Express. For Rolls–Royce, this is one program among many; for BMW, it is the entire business.

How can BMW and Daimler–Benz, two successful automobile companies, look at the aircraft engine market and reach apparently opposite conclusions? At Daimler–Benz, Edzard Reuter's grand diversification strategy failed to succeed and probably has fallen into disrepute. A cynic might suspect that BMW has yet to encounter such a failure and, therefore, continues to hold to its optimistic plan. Perhaps also the Federal Ministry of Economics, which claims some oversight of the industry, encouraged BMW to set up shop in the former East Germany (MTU has also taken over an engine overhaul shop at Pirna and is managing it as part of MTU Maintenance AG, its own commercial engine overhaul activity in Hanover) and then encouraged BMW and MTU to consolidate their aircraft engine activities.

Chapter 20

Product Diversification

In THE business of making aircraft gas turbines, what other options are available for increasing revenue? Management consultants usually advise their client companies to focus management attention and other resources on the core product. They frequently also look for other applications for their clients' products or core competence and are capable of defending both strategies with equal conviction. Each can be successful: a strict focus on the core business avoids frittering away resources on less-profitable opportunities; additional markets for core competencies can be successful if the incremental profits exceed the incremental costs. To GE, warship propulsion appeared to offer such an opportunity.

In the late 1950s, the British navy ordered some fast patrol boats driven by Bristol Proteus turboprop engines. British patrol boats and torpedo boats had used modified Rolls–Royce Merlin piston aircraft engines during World War II (German patrol boats, the so-called E-boats and S-boats, used high-speed/lightweight diesel engines. The German Navy continues to use such engines to this day.) But gas turbines offer higher power in less space and for less weight—what the naval architects call greater power density. That is important in these coastal warships, whose mission consists of a high-speed dash toward the target and a high-speed escape after they launch their torpedoes. At about the same time, high-speed hydrofoil boats were being developed in the USSR and Switzerland to serve as passenger ferries. Such boats ride up on a submerged wing or hydrofoil once they exceed a certain speed. The foil lifts the weight of the hull completely out of the water and reduces the drag of the normally displaced hull. Such boats, whose attraction is their high speed while foilborne, also need engines of high power, compact size, and low weight.

In 1959, hydrofoil boat development in the United States was sponsored by the U.S. Navy. Boeing Aircraft Company, located in Seattle, where ferries connect the city with many of the islands in Puget Sound, designed a hydrofoil boat propelled by a 500-hp gas turbine which Boeing developed originally as a truck engine. The Boeing 501 design was later licensed by Boeing to Fabrique Nationale in Belgium, which sold a number of the gas turbines to military customers for aircraft engine ground starters and auxiliary electric power generation, and to the Swedish Army to install in its S tank as a compact boost powerplant supplementing the basic diesel propulsion engine. The 500 hp available almost instantly from the gas turbine allowed the S tank to sprint out of its firing position within seconds after firing its gun—out of the way of return fire.

On Long Island Sound, Grumman Aircraft developed the HS Denison, a hydrofoil boat displacing about 100 tons, on the scale of a torpedo boat, requiring much more installed power than the Boeing boat. GE designed a power turbine for the J79 gas generator and created an engine of some 15,000 hp, the LM1500 (LM was used by GE to denote "Land/Marine;" 1500 was simply the maximum power divided by 10). Installed in the Denison, the engine enabled it to reach a top speed of more than 60 kn. The LM1500 was also installed in an even larger hydrofoil, the 300-ton *Plainview*, and as a high-speed boost engine in an experimental U.S. Coast Guard cutter, the *Hamilton*, which had a normal displacement hull.

Nothing much came of these experiments. The development work was done by two aircraft companies on Puget Sound and Long Island Sound, sheltered bodies of water with normally smooth surfaces, but the U.S. Navy and the U.S. Coast Guard operate in "blue water," out of the shelter of land. Operating conditions there are much harsher, and small ships of 100– to 300-ton displacement lack the endurance for typical "blue water" missions. Although U.S. Navy interest in hydrofoils persisted for some time, culminating in the gas-turbine driven Pegasus class some 15 years later (propelled by the GE LM2500 gas turbine, a 25,000-hp derivative of the TF39 turbofan), even these found limited application. Interdiction of drug runners in the Caribbean could be handled more economically by conventional ships carrying patrol helicopters. Such ships had greater endurance for patrol missions, and space for more comfortable quarters for the crew.

For GE, the LM1500 program had involved design and development but little production. Suddenly a new market niche appeared. The U.S. Marine Corps was looking for ways of launching aircraft from very short runways on or near assault beaches, in principle a portable catapult similar to that on the aircraft carrier from which their attack aircraft would transfer to the airfields in the battle zone. All-American Engineering Company designed a portable catapult to go with the aluminum mat combat runways, needing about 15,000 hp. The arrester gear for landing was an easier engineering problem.

GE had just completed development of the J79-2 engine for the U.S. Navy's F4H Phantom carrier-borne fighter. The engine needed many refinements to improve its reliability and suitability for service in a naval environment, because aircraft carriers have limited hangar space for maintenance and repair and for spare engines. With much prodding from the U.S. Navy, GE developed the J79-8 version, which became the definitive powerplant for the F-4B version of the Phantom (a sister version, the J79-15, was used in the U.S. Air Force's F-4C and another, the J79-19, in the single-engine Italian Air Force F-104S and the Israel Air Force Kfir, a derivative of the Mirage III), and the U.S. Navy found itself with some 50 J79-2 engines left over. The U.S. Navy gave GE a contract to convert these J79-2 engines into LM1500 gas turbines to be used in portable catapults for the so-called short airfield for tactical support, or SATS.

A few LM1500 engines found commercial applications, to drive compressors for oil and gas pipelines. Most compressor stations are remote from major towns and there is some advantage to the power density of the gas turbine. It is easier to transport to a remote site and easier to set in place than a much bulkier diesel engine of the same horsepower, and it can burn some of the natural gas tapped from the pipeline as fuel, obviating the need for a large tank for diesel fuel. In such applications, low engine cost is important: the benchmark is the price of a diesel engine of comparable horsepower. Set against these successes, one LM1500 application was a triumph of misguided engineering optimism. The Cincinnati Gas & Electric Company needed to add capacity to generate electric power at times of peak demand on its Middletown plant, not far from GE's Evendale factory. Such a peaking powerplant need not have the highest efficiency, but it must be less expensive to buy per kilowatt of installed power than the steam turbine generators used for base load, and it must be capable of coming up to full load in a matter of minutes. Middletown and GE conceived a 150,000-hp engine consisting of a huge turbine wheel driven by the hot exhaust gas of 10 LM1500 gas generators. The 10 gas turbines were arrayed in front of the power turbine wheel like chambers in the cylinder of a giant revolver. There is an old engineering prejudice that the best designs are the simplest and should look elegant as well as operate efficiently. That does not always hold true, but it must have been so in this case. This was not the best way to generate peaking power. In any case, it faced a formidable competitor. GE's Power Generation Division, GE's traditional core business and one of the biggest divisions in the company, regarded electric power generation as its exclusive responsibility. The steam turbine people protested to GE's management that the aircraft gas turbine people ought to stick to aviation and leave the electric utility companies to those parts of GE with experience serving that market. Corporate headquarters agreed.

Several industrial applications were also found for smaller aircraft gas turbines. The Unit Rig Company of Tulsa, Oklahoma, makes huge ore trucks for open pit mines, with a capacity for 100 or 150 tons of ore. Each wheel of such a truck is driven by a 250-hp electric motor in its hub, and a 1000-hp diesel engine in the truck drives a generator that furnishes power to the electric motors. Unit Rig Company hoped that a small gas turbine might take up less space and weight on the truck. GE proposed a version of the T58 helicopter engine, the LM100, rated at 1000 hp. The LM100 was in fact much smaller and lighter than the diesel it replaced, but it had a much higher fuel consumption, so that the combined weight and volume of engine and fuel was about the same as that of the diesel. And the

LM100 cost much more than the diesel, whose manufacturing cost had been reduced over the production of thousands of such engines. Abrasive iron ore dust in the mine also caused a rapid deterioration of the efficiency of the LM100 compressor, whose blades are so small that even a little abrasion had an immediate effect. GE and Unit Rig Company argued that the LM100 used less fuel than the diesel in the actual operating environment of the Mesabi iron mines in Minnesota, where the diesel had to be kept running at idle power all night long to keep it from freezing; whereas the gas turbine had been designed to be capable of starting at temperatures as low as $-65°$ F. But tortured rationalizations do not impress customers. Unit Rig Company quickly reverted to diesel power for its trucks.

Such problems can find their solution when the incentive is great enough, for example, in tank propulsion. The engine compartment and fuel tanks of a tank must be armored: the bigger they are, the heavier the tank. For the M1 battle tank, the U.S. Army selected a 1500-hp gas turbine developed by Lycoming, the AGT1500, because the combined volume of gas turbine and fuel tanks was smaller than that of a 1500-hp diesel and its fuel tanks. The AGT1500 has a recuperator to improve its fuel consumption and an elaborate air filter to keep out abrasive dust. In the Gulf War of 1990, M1 tanks and their engines performed well. In contrast, the German and other armies have spent a lot of money on the development of compact 1500-hp diesel engines, which become much more complicated than conventional diesels and, therefore, less reliable and more expensive.

Some intriguing applications must be discouraged. A race-car builder bought a GE T58 helicopter engine from a dealer in surplus military equipment and installed it in an old-fashioned Indianapolis race car chassis for the Indianapolis 500. The Indianapolis technical committee was horrified at the competitive threat posed by a 1200-hp engine of a completely new type—typical Indianapolis car piston engines at that time developed 600 hp. The committee decided to deal with the competitive threat to their established engines by setting up restrictive rules for gas turbines: the airflow annulus area could not exceed a certain value, smaller of course than that of the T58. Nothing loath, the racer put the compressor of his T58 in a lathe and ground off 0.5 in. from the tips of the first stage of compressor blades to match the new specification. He reassembled his engine, installed it in his car, and ran it around the Indianapolis track, scaring the hell out of all of the other drivers. Somewhere along the way he asked his cousin, a general in the U.S. Air Force, to find out from GE whether his light-hearted modification would hurt the engine.

Engineers love problems, and most of those at GE would have relished working on race cars. But Gerhard Neumann decided that there was considerable liability risk in this, risk of an accident in which the driver, other drivers, and spectators could be injured. And there was no reward that would justify the risk. GE told the car owner that there was some risk of compressor failure, that GE advised strongly against running the car in the race, and that it would have nothing to do with modifying the engine. In any event, the car did not qualify: the driver could go very fast on the straight sections of the track but was unable to slow down entering the two curved sections, where piston engines rely on compression braking to supplement their friction brakes.

Finally, there was a handful of LM100 and LM1500 installations in Japan for earthquake emergency electrical power generation in key urban office buildings, such as telephone central exchanges, and one experimental torpedo boat propelled by three modified T64 engines, but these were more demonstrations of the marketing power of IHI, the company building the aircraft engines under GE's license, than evidence of a market big enough to justify the technical effort.

All of these marine or industrial installations involved engines modified from aircraft gas turbines GE had in production. They were considered incremental business, with the additional sales justifying the modest development cost for the conversion. There was, however, one small project outside the core business of the GE aircraft engine business that violated this conservative strategy. It involved designing a small gas turbine directly for industrial applications and it was not a success. In the late 1950s, Gene Firestone was the head of the Large Jet Engine Department in Evendale and one of the principals negotiating the J79 license agreements for the F-104 Starfighter program in Europe. In the course of discussions about the J79 license agreement, Firestone visited the KHD factory in Oberursel, near Frankfurt am Main. The German government planned that some of the J79 parts would be manufactured by KHD, a large engineering company known for its air-cooled truck diesel engines and as a supplier of diesel engines to the German federal railway for high-speed passenger locomotives. Klockner–Humboldt–Deutz wanted to add a gas turbine to such locomotives for acceleration boost power. Firestone felt that GE could easily design such a gas turbine, given that it would be free of the stringent weight limitations of an aircraft engine and would operate under much simpler conditions. He set up a team in Evendale to design a simple 500-hp gas turbine, called the LM50. Its chief designer had been hired by GE from Solar Gas Turbines in San Diego, California, which was designing a series of small gas turbines as mobile auxiliary power units and engine starters for the U.S. Air Force and which later designed and built stationary gas turbines with ratings up to 5000-hp for industrial applications.

The LM50 ran—not very well, but it ran. GE began to reflect on its future. Could a small gas turbine be produced at a cost competitive with a diesel engine of the same rating? There were many on the market already. Was GE better than Solar Gas Turbines at designing this size of machine? What was the market for such an engine? How many locomotive boost engines would the German Bundesbahn need? Would the market justify the investment in engineering and product support? It did not take long for GE to conclude that it would gain more from devoting its resources to the development of large aircraft engines than it would from playing around with a 500-hp auxiliary engines for the railroad. GE bundled up two sets of LM50 parts and two sets of drawings and sold them and all the rights to the LM50 to KHD.

But there was an unintended negative consequence: the parts were for development prototype engines and not of the finish and quality expected by KHD. The German company complained to the government that GE had palmed off its junk for a high price. In the case of other quality complaints at the time, GE was able to furnish replacement parts at no charge and repair its reputation; that was not possible with the LM50 because there were no other parts.

Soon afterward, KHD began to manufacture 1000-hp T53 helicopter engines under license from Lycoming and installed one in a Bundesbahn locomotive. But like GE's LM100, the T53 was not really competitive in cost with a diesel engine of the same power. Klockner–Humboldt–Deutz decided to concentrate its gas turbine work on military aviation applications, where the market was more certain and the pressure on price less, and continued in the aircraft gas turbine business with research funding from the German government. In partnership with SNECMA, KHD subsequently designed and built the Larzac, a small turbofan engine for the Franco–German Alphajet trainer, followed by a gas turbine starter and auxiliary power unit for the Tornado strike fighter. Both products were complex and expensive, but the costs were underwritten by the German government. When production of the Alphajet and Tornado came to an end, the KHD gas turbine division ran out of work and was put up for sale. A buyer was found: BMW, which had gone back into the aircraft engine business and needed production capacity for its share of Rolls–Royce Tay production and for the new BMW/Rolls–Royce 700 engine.

As these stories illustrate, no market of any attractive size developed for industrial gas turbines derived from aircraft engines. The conventional wisdom was that aircraft derivatives were too expensive to be competitive with other prime movers and, being perforce birdlike, were too flimsy and, therefore, unreliable. They could not compete on manufacturing cost with diesel engines at ratings up to 5000 hp or with steam turbines for the higher power ratings required for electric power generation. Aeroderivative gas turbines had the advantage of high power density, important for pipeline compressors in remote installations, but only Rolls–Royce had managed to solve the cost problem by buying several hundred Avon engines from the RAF as surplus, and then modifying them into industrial engines for pipeline pumping. The cheap gas generators made the gas turbine price competitive.

In the mid-1960s, the U.S. Navy began studies for a new ship, the DDX class, with a displacement of less than 10,000 tons and 80,000 hp installed, to replace destroyers and cruisers in the fleet. Propulsion system volume had to be low and capable of being distributed in the hull so that no one hit would immobilize the ship, while both its weight and volume had to be kept low to maximize the space available for the ship's new electronic systems, weapons, and crew. The weight and volume of fuel would be counted as part of the propulsion system. The Navy considered three different propulsion systems: steam turbines operating on superheated steam at a pressure of 1200 psi, gas turbines designed along conventional steam turbine practice, and gas turbines derived from aircraft engines.

GE has a long history as a supplier to the U.S. Navy: steam turbines and generators for electric ship propulsion, control systems for gun turrets, engines for torpedoes, and propeller speed reduction gears for warships, particularly for quiet nuclear submarines. GE decided to make a bid for the propulsion system for the new class of ships and proposed three entirely different systems. Two were from GE Power Systems Division: 1) in conjunction with the boiler engineers Babcock & Wilcox, a high-pressure steam turbine system derived from that in the Knox class of destroyers that the DDX would replace, and 2) a gas turbine system designed in conformity with steam turbine practice, with large, heavy castings deemed sturdy and reliable enough for marine service. The Flight Propulsion Division proposed a third system, four lightweight 25,000-hp gas turbines derived from the TF39 turbofan, driving variable-pitch propellers.

The Seapower Subcommittee of the U.S. House Armed Services Committee invited GE to explain its proposals. Each division sent its vice president of engineering to present his case, with Fred W. Garry representing Flight Propulsion. The committee chairman asked him how GE could defend three different proposals with apparently equal conviction, to which Garry replied that each system had its merits and that, in the end, the U.S. Navy would select the system that best matched its overall needs. GE was convinced that the customer could make a better decision than GE would have done by concentrating on one approach or the other.

The Navy's technical evaluation showed that the traditional steam turbine system was not as reliable as conventional wisdom had suggested. At the very high steam pressures required, the boiler tubes and steam piping were the sources of frequent steam leaks. These were of course dangerous to the crew, and they also significantly degraded the available power of the steam propulsion system. The U.S. Navy chose gas turbines and specifically aeroderivative gas turbines. But their reliability had to be proven.

In 1967, the U.S. Navy launched a high-speed cargo ship, the gas turbine ship (GTS) *Callaghan*. The only 20,000-hp gas turbine engines available were marine conversions of the Pratt & Whitney JT4D commercial aircraft engine, and two of them were installed in the *Callaghan*, which sailed on weekly liner service between Bayonne, New Jersey, and Bremerhaven in Germany, carrying vehicles and other cargo for the U.S. Army in Europe. The objective of the FT4 installation was to validate the usefulness of aeroderivative gas turbines for high-speed ship propulsion, and the FT4 performed satisfactorily. Meanwhile, GE created a task force to develop a marine derivative of the TF39, the LM2500, rated at 25,000 hp, by removing the fan from the front of the engine and using the fan turbine to drive into the ship's gearbox out of the rear of the engine. In March 1968, GE proposed that the U.S. Navy install one LM2500 in the GTS *Callaghan*, leaving one of the Pratt & Whitney FT4 engines in place, to get a direct comparison of the two designs. The U.S. Navy accepted, and set a target date of late 1969.

GE looked for a partner to invest in the development of the LM2500. GE's vice chairman Jack Parker had previously met Giovanni Agnelli, the chairman and biggest shareholder of FIAT, to discuss opportunities for the two companies to collaborate. FIAT was already building J47, J79, and T64 engines under GE license and had shared with Alfa Romeo in T58 and J85 production. Parker suggested that FIAT participate in the development and production of the LM2500, with GE responsible for the overall system design and the gas generator, a derivative of the TF39. FIAT would design all the parts unique to the LM2500, such as the new structural frames and auxiliary systems, and would test the complete engine in a new test cell with a 30,000-hp water brake. In production, FIAT would build all the parts unique to the LM2500 engine, GE all the parts common to the TF39. In addition to production for GE, for its investment of about $10 million FIAT would also have a license to assemble and sell LM2500 engines on its own behalf. Parker and

Agnetti shook hands, and Parker added, "Now let's see how long it takes the ribbon clerks to screw it up."

This sensible co-production arrangement almost caused the deal to come apart because GE's lawyers negotiating the details were sure that the U.S. Navy would insist on retaining the right to buy all parts of the system through competitive bids and would not accept FIAT as the sole source for the unique parts. But there was no need to worry, GE assured FIAT; because FIAT would be the initial source for the parts, it would always have a cost advantage over any later parts competitor. FIAT was not at all reassured that it would get fair treatment from GE purchasing agents or the U.S. Navy, and perhaps not confident that it could be cost-competitive. But the two managements were able to agree that they had to bring about what Parker and Agnelli had shaken hands on. Ambiguous words were found that satisfied both FIAT and the GE lawyers, and the technical work began at high speed. A combined FIAT/GE team in Turin and Evendale scrambled to design the engine and build the new test cell in Brindisi, with a German water brake (a water brake is a test device like a dynamometer that applies a realistic load to the output shaft of the engine on test by means of water pumps, and dissipates the engine's energy output in the form of heat) of a size larger than any on the market. Parts were made and the engine finally assembled and tested, meeting specifications. In late 1969, this first LM2500 engine was installed in the *Callaghan*, where its performance quickly eclipsed that of the FT4 on the other shaft. The higher pressure ratio and the higher turbine inlet temperature of the LM2500 resulted in a significantly lower fuel consumption than that of the FT4. However, GE quickly learned a basic chemistry lesson: salt air corrodes turbine parts at a rate unknown to aircraft engines. There was a serious degradation of performance after some 2000 hours of operation, but GE's metallurgists at Evendale and the corporate research and development center quickly developed corrosion-resistant coatings that solved the problem. After some 25,000 hours of successful LM2500 operation (more than four years of liner service back and forth across the Atlantic Ocean), the remaining FT4 was replaced by a second LM2500. The difference in fuel consumption was a triumph, and successful operation of the engines demonstrated to the U.S. Navy that aeroderivatives were at least as reliable and long-lived as any other propulsion system. To its delight, the U.S. Navy found an unexpected benefit: aeroderivative turbine engines could be brought up to full power from a dead start in less than 1 minute; it takes much longer to build up steam pressure from a cold start and bring turbines up to operating temperature.

When development of the LM2500 began in earnest, Neumann assigned program management responsibility to Sam Levine, a manager with years of experience in the design and development of aircraft gas turbines. Levine also had the task of convincing the U.S. Navy and the shipyards that would build the new DDX that the LM2500 was the best engine for the application. Levine drove his American–Italian team hard and successfully. In December 1970, Litton Shipyards, the leading DDX designer, recommended the LM2500 to the U.S. Navy, which accepted the recommendation and awarded GE a contract for 120 engines for the largest naval surface ship-building program since the end of World War II (the ships were designated the DD-963 Spruance class after the lead ship). In 1973, a manufacturing specialist took over program management from Levine, his task to manufacture the LM2500 at a cost lower than the contract price Levine had signed with the U.S. Navy.

In 1972, the U.S. Navy launched another similar class of ships, the FFG-7 *Oliver Hazard Perry* patrol frigate and again selected the LM2500 for this class, this time two engines per ship, driving a single propeller. Other navies followed, first the Italian navy, for which FIAT built LM2500 engines for frigates and a helicopter carrier, then the navies of Germany, Spain, Korea, Indonesia, and Denmark, whose shipyards designed frigates and large patrol boats using the LM2500. The most unusual sales were for four DD-993 frigates ordered by the Shah of Iran for his navy—these were taken over by the U.S. Navy after the Iranian revolution in 1978—and engines for Chinese frigates in 1979.

The USSR adopted gas turbine power for its warships more quickly than the U.S. Navy. Many of the Soviet gas turbines were of the heavy-duty type and some were aeroderivative. One Nikolaev gas turbine obviated the need for reversing gear or a reversible-pitch propeller by a valve ducting hot gas to a reversing turbine whose blades were mounted on the tips of the power turbine used for steaming ahead, a curious and complicated mechanism, but one that must have worked reasonably well. Other than the Soviets, only Rolls–Royce offered any significant competition to the LM2500 for warship propulsion. The British and Dutch navies and the Japanese Maritime Self-Defense Force used Rolls–Royce gas turbines.

After its success in warship propulsion, the LM2500 found two other applications. The first was electrical power generation (the power turbine is designed to rotate at 3600 rpm, ideal for 60-Hz electrical generators) and compressor drives on oil- and gas-drilling platforms. As huge as such platforms are, each pound of weight on the platform means tons of steel in the supporting legs. The aeroderivative gas turbine has a unique advantage here of lower weight than diesel engines or heavy-duty gas turbines of the same power rating, and the LM2500 found a ready market on North Sea platforms and elsewhere. A second similar market developed in pipeline compressor and pumping stations, where LM2500 engines have operated for as long as 20,000 continuous hours on natural gas fuel, with no need for maintenance. There was strong competition for pipeline pumper gas turbines from the manufacturers of so-called heavy-duty gas turbines, whose mechanical design is derived from steam turbine practice. Their name and massive cast casings endow heavy-duty gas turbines with a reassuring aura of reliability and long life, even though those are clearly matched by the refined engineering of aeroderivative gas turbines. Most of the world's manufacturers of steam turbines offered heavy-duty gas turbines, and their leading producer was GE's own Power Generation Group, by no coincidence the world's largest manufacturer of steam turbines. GE's heavy-duty gas turbines competed fiercely with the LM2500 for pipeline applications. Between the two types of gas turbines, GE held the world's largest market share.

Only in base-load electrical power generation did the LM2500 fail to find a significant market. The reason was its size. Even for peaking power, utility companies prefer power in blocks of 100 MW or more and, for base load generation, they want 300-MW units or bigger; at 20 MW, the LM2500 was too small to justify its price. To get more power, GE substituted the gas generator of the CF6-50 for that of the TF39, thus producing the LM5000 industrial engine. With its greater airflow, the LM5000 was rated at 30 MW and found a small market in special circumstances, for example, power generation at a handful of large chemical plants in California that could use the exhaust heat for chemical processes and sell excess electric power into the utility grid; another specialized application was a portable electric powerplant floated on barges from Japan to its site in Bangladesh. In contrast to the LM2500, stationary gas turbines could be designed along heavy-duty lines for ratings of 90,000 MW or more to meet the needs of the base-load market. GE Power Systems did so, as did Siemens, ASEA–Braun–Boveri, and all the other steam turbine manufacturers. GE's aircraft engine specialists contributed to the power systems developments in the design of combustion chambers and the aerodynamic and mechanical design of compressor and turbine blades.

The market for aeroderivative gas turbines was wherever high power density was important, as in the above examples, where other forms of prime movers were too big, too heavy, and often unreliable. Where volume and weight were not important, for example, for the propulsion machinery of very large cargo ships and oil tankers, other forms of prime movers were usually lower in cost and sometimes (but not always) in fuel consumption. For electric power generation, the steam turbine manufacturers continue to dominate the market, along with industrial gas turbines of a power output higher than that available from existing aircraft engines.

Thanks to the U.S. Navy's launch order, the LM2500 was a profitable business for GE and FIAT, which have built about 1700 LM2500 gas turbine engines. The other GE aeroderivative programs were probably more distractions than profitable business opportunities.

Chapter 21

Service as a Separate Business

MARINE and industrial gas turbines illustrate how engine companies can look for other markets and applications for their products and they can do the same in services. Here are the fundamentals of gas turbine service: aircraft engines require scheduled inspection and maintenance to keep them flying safely and efficiently. Sometimes they need unscheduled maintenance, when an inspection reveals a problem of wear or failure in a part, or if a problem has occurred in operation. Operators keep track of trends, in parameters such as exhaust gas temperature, vibration, or metallic particle levels in the engine oil, as a barometer of impending trouble. The ideal is never to have a part fail during its intended service life, a life long enough to provide a low cost per hour and to replace parts just before they are about to fail. When parts have been removed from an engine for wear, airlines want to be able to restore them to a serviceable condition, if possible, with a further life span equal to the first installment.

Who does the maintenance? Who repairs the parts removed after a scheduled inspection? There are several possible answers. Large operators, such as most air forces and major airlines, have their own overhaul and repair operations, staffed with their own engineers and technicians, and believe that they know more about the engine in service than the designers. The operators may call on the engine manufacturer for technical analysis and advice, but they will then often set their own inspection intervals, develop their own repair procedures, and test them to demonstrate to government airworthiness authorities such as the FAA that the repair is safe and effective. In the spirit of maintaining the overall level of flight safety, they usually share their technical findings with the engine manufacturer and other users of the same engine model in return for the experience of the others.

Smaller air forces and smaller airlines may not have the same technical capability, but they turn instead to larger operators for maintenance and repair services. The U.S. Air Force has done maintenance for years for foreign air forces that use American military aircraft. Major airlines such as United Airlines or Lufthansa, with strong engineering departments, will do contract maintenance for other smaller airlines.

There are independent maintenance companies that offer such service to any airline or air force that does not do such work itself. Like the major airlines, these firms develop their own repair procedures and have them validated through service tests by their customers. There are several such commercial maintenance companies in the United States and Europe as well as in East Asia, some of them government owned. When a government decides that it wants to have its own national aircraft engine capability, overhaul is often the first step. Designing engines requires a huge investment in money, people, facilities, and time, beyond the resources of most governments. Even making engine parts requires a high level of technical skill and a large investment in tooling and machinery. Japan, one of the few nations that elected to make the investment required for research and development and manufacture, has the resources to succeed and the patience to wait for success; even so, the results have been equivocal, as we have seen. Setting up a company to overhaul engines and repair parts for the local air force requires a much smaller investment and has the added advantage that it will then also be capable of assembling and testing production engines. Examples of such companies are Hellenic Aerospace Industries in Greece; the Turkish Air Force engine overhaul depot at Eskisehir; and the Universal Technical Maintenance Center in Bandung, Indonesia. Typically, governments rely on the engine designer for technical assistance to set up such companies and sometimes even to manage them.

The designer and manufacturer of the engine also offers engine overhaul and component repair. In the former USSR, the factory was the only source for this, forcing Soviet operators to carry large inventories of spare engines. In the West, Rolls–Royce, Pratt & Whitney, and GE all offer overhaul and component repair to their military and airline customers to supplement the operators' own capacity and that of independent overhaul companies. The manufacturer not only knows more about the product than anyone else but also has a direct interest in keeping the engine running to the satisfaction of its customers, who expect the manufacturer to collect and analyze reports from all operators about problems with the engine in service, to be ready to overhaul their engines for them, and to repair their parts—as well as giving them technical assistance whenever they elect to do such work themselves, for themselves or for third-party customers. In effect, the engine manufacturer is expected to support his customers with overhaul and teach them how to compete against him when they choose to do the work themselves or have it done by independent overhaul shops.

The cost-accounting of all prime engine companies allocates the overhead expenses of engineering, project management, marketing, and product support to each labor hour of work done for external customers. Such overhead costs are high, much higher than the overhead costs of independent overhaul contractors even when the direct labor hours for a repair task are the same. The cost of overhaul and repair by the original manufacturer is often high enough to justify an airline's investment in its own overhaul facility—and usually much higher than the costs of independent overhaul contractors even fully burdened by their own overheads. Furthermore, the desire of governments, for example, in Taiwan, Greece, Indonesia, to set up their own overhaul shops, with the investment written off against national military programs, has created a surplus of overhaul capacity and pressure on overhaul prices. It is difficult to be profitable in the engine overhaul business, particularly for an engine manufacturer with its big overhead expenses.

GE initially looked at engine overhaul as an ancillary service, a mechanism required to support the marketing of commercial aircraft engines: the service must be offered because the customers demand it. To minimize its cost in the face of price competition from the independents, GE set up the aviation service department (ASD) as a completely separate organization, with its own overhead expenses—lower than those of the parent organization—to furnish technical support and to develop and validate repair processes. As time went on, GE assigned other tasks to ASD. With its low overhead, it was able to produce maintenance tooling at a competitive price for airline and military operators, some of whom used to demand tool designs from GE so that they could make the tools themselves. When GE customers decided to set up their own overhaul facilities, ASD put together teams to design them for the customer, wrote specifications for the machinery and purchased it for the customer, and trained the customer's technicians. In some cases, customers contracted with ASD for a management team to operate the new facility until their own managers had adequate experience.

Because such customers then compete with GE for overhaul work, there are inherent contradictions in all of this. Air forces are run by soldiers and defense ministers, seldom by economists or businessmen. Some found to their surprise that once their new overhaul depot had taken care of the needs of their own air force there was not enough work to keep all of the investment loaded to capacity. They appealed to GE to help them market their excess capacity, a difficult task because GE expected to face the same demand from other customers in the same situation. Some customers went so far as to denounce GE for staying in the overhaul business at all, because that was competition for the overhaul work they had hoped to get from third parties. GE had already concluded that there was little profit in the overhaul business because of the excess of capacity available worldwide and the resulting heavy pressure on prices. Nevertheless, as the prime designer of the engine, GE felt that it had to offer overhaul service. Some customer airlines had no capability of their own and, fearing that their needs would always take second place to those of the owners, did not want to buy the service from competing airlines or independent shops. For such customers, GE therefore offered a minimum of overhaul capacity but did not want to go beyond this level, and instead concentrated on the repair of engine parts removed during overhauls, such as the refurbishing of combustion chambers, turbine blades, and vanes after thousands of hours of operation at very high temperatures and the refurbishing and polishing of fan and compressor blades and vanes damaged by stones and other debris sucked into the engine. One interesting repair technique adds metal to a worn part, for example, by electron-beam welding a tip onto a worn blade or plasma-spraying a coating onto a worn bearing surface. The added metal is then heat treated and finish machined to the original contour. Such work requires the highest precision. Because of its background in the design of the part, GE had a technical edge in such repair work, which it was able to turn into a competitive advantage against independent shops or the airlines themselves. Parts repair became a profitable business even as overhaul itself became less so.

Given the excess capacity in the overhaul business and the inevitable tensions of competition between the manufacturer and its own customers, how did GE find itself the owner of British Airways' engine overhaul shop in Cardiff, Wales? In 1990, airlines began to order the Boeing 777 for their high-density routes. As usual, all three major engine companies developed new engines for this big twin. Pratt & Whitney offered a growth version of the PW4000, Rolls–Royce a growth version of the Trent, and GE a newly developed engine, the GE90. And, as usual, the competition for launching airlines was intense: the manufacturers all believe that it pays to be first, that it pays to win at the major airlines, which are usually the first to order a new aircraft type.

United Airlines was the first 777 customer and picked the Pratt & Whitney engine. United Airlines has been a devoted user of Pratt & Whitney engines, although it has used other makes when they offered significant advantages. Then came British Airways, for years a user of Rolls–Royce engines. When both companies were owned by the British government, it often seemed to Americans that British Airways was a captive customer, although that overlooks the fact that Rolls–Royce offered engines of the highest performance and reliability, fully competitive with American engines; small differences in specific performance can always be made equal with price concessions.

Brian Rowe, then CEO of GE Aircraft Engines, an Englishman by birth who received his first practical training as an engineering apprentice at DeHavilland Engines, made it his personal business to cultivate the management of British Airways. During the days when GE had a cooperative agreement with Rolls–Royce on the CF6-80C and the RB.211-535, Rowe also made a special effort to meet the British civil servants and politicians who were responsible for the aviation industry and aeronautical research and development. When the RAF terminated the unsuccessful development of the radar for the Nimrod airborne early warning aircraft and ordered the Boeing E-3B with CFM56 engines, GE Aircraft Engines placed offsetting orders for engine parts with suppliers in the United Kingdom and set up an engineering office in Leicester, not far from Rolls–Royce. One purpose of that was to create a site for British engineers hired by GE to do productive work until visas and work permits could be arranged for them in the United States, but another objective was to draw on the unique technical capability of a British company, GEC Limited (despite the confusing similarity of names, there is no corporate connection between the American firm GE and the British firm GEC Limited), to develop the LM5000 and other advanced industrial gas turbines. GE was able to demonstrate convincingly that its programs contributed significant engineering and production workload to the British engine industry.

The British properly regard Rolls–Royce as a jewel in their industry's crown—an engineering firm with large exports, whose products are among the best in the world. But there are some experts who worry that the smallest of three competitors in a fiercely competitive market will have difficulty surviving. The market pays no premium for Rolls–Royce engines when they are the equal of Pratt & Whitney or GE engines. Rolls–Royce's costs are likely to be higher if its sales are significantly smaller, leaving less money for development of new and more advanced products, potentially a vicious circle. Rolls–Royce has a monopoly over Britain's military engine business, and particularly over the military spare parts business. No doubt these create a steady cash flow, but this may perhaps not be enough to maintain the highest level of investment for commercial engines. There were knowledgeable people in the United Kingdom who believed that Rolls–Royce needed to ally itself with either GE or Pratt & Whitney to survive.

The competition for engines for the 777 was a new kind of challenge for GE, and for Rolls–Royce and British Airways. Although the government retained enough interest in each to protect the national interest, the two British companies had been privatized and were free to invest without regard to political considerations. The government was keenly aware of the fact that both would have to compete on a global basis, with no leeway for anything but the highest technical and economic performance. GE, the foreign competitor, brought work into Britain by placing significant orders of high quality with British manufacturers and engineering firms and had cultivated the British Airways decision makers as well as the politicians who had a concern about the outcome. British Airways chose the GE90 for its Boeing 777s, the first customer for the new engine. There was a moment of anxiety when GE was late certifying the engine against damage to the fan blades from bird strikes (a similar problem had undone the original RB.211 development for the Lockheed TriStar and had contributed to forcing Rolls–Royce into receivership in 1970), but the problem was resolved in three months and the engine went into service with excellent results.

There was outrage in Britain, of course, at the choice of other than a Rolls–Royce engine, but it subsided quickly when British Airways made the public argument that it had to be free to select the product that best allowed it to compete on the world market. Since privatization, British Airways had taken drastic steps to improve its productivity, offer better service, and reduce its costs, ruthlessly pruning any activity that did not contribute to these goals. Secondary services were sold, and some work previously done by the airline was farmed out to more cost-effective subcontractors. Among these secondary activities was the airline's big engine overhaul facility in Cardiff. British Airways' core business was moving passengers from airport to airport at a profit, and it decided that it ought to be able to buy engine overhaul at a better price than the cost of maintaining its own dedicated engine overhaul shop. British Airways suggested that GE buy the Cardiff facility and that, if GE did so, British Airways would certainly consider the shop for its overhauls if the price were competitive. GE agreed and is reported to have paid some $350 million for the entire facility with its skilled work force. Was this a condition for British Airways' choosing the GE90? Was it a disguised price concession on the engines? There is no way of separating the transactions so that GE and its competitors are free to construe it as they wish. But there is no doubt that GE treats the Cardiff overhaul shop as a profit center rather than as an adjunct to facilitate marketing. GE is now offering overhaul there to airline and military customers, not only for the GE90, but also for Rolls–Royce and Pratt & Whitney engines, expanding the product scope of the original British Airways shop. Including Cardiff but without considering the sale of spare parts, the entire overhaul and repair revenue of GE Aircraft Engines must be close to $1 billion per year—of the same order of magnitude as GE's sales of military engines to the U.S. government (these hit a peak of about $3 billion in 1986 and have declined since then with the rest of the U.S. DOD procurement budget).

Chapter 22

The Death of Douglas

DOUGLAS Aircraft's history began in 1932 with the DC-1, a twin-engine low-wing metal airliner with 20 passenger seats, which evolved quickly into the DC-3, the world's first universally successful airliner. Out of that came a series of other successes: first the DC-4, soon followed by the widely used DC-6 and DC-7. Their main competitor was the Lockheed Constellation, against which Douglas managed to hold on to its leading market position. Boeing was quicker into the market of jet-powered airliners with the 707, but even then the DC-8 with its great range and efficient wing managed to gain a respectable market share. The twin-engine DC-9 was another best seller in the short- and medium-range market.

But from then on it was all downhill. The new DC-10 cut short DC-8 sales and had to fight the Lockheed L-1011 Tristar as well as the Boeing 747 for every sale of its own. When Lockheed dropped out, Airbus Industrie came in, offering attractive new airliners incorporating the latest technology, and again every sale became a price competition. Douglas was clearly not generating enough profits to justify investment for developing a new generation of airliners and was taken over by McDonnell Aircraft, which had a large and profitable military aircraft business. McDonnell was unwilling to invest in the development of a new airliner, presumably because the return on such an investment would not be attractive. Airbus Industrie was able to raise money with the aid of the four governments that had supported its founding, while Boeing could draw on the monopoly profits of its 747 to develop new products. When the demand for airliners shrank in the early 1990s, Douglas gave up the most market share and ultimately became irrelevant in the competition for new airliners. In the end, Boeing took it over in 1997, keeping McDonnell's lucrative military business but shutting down much of the Douglas airliner business, except for two models for which there was still some demand.

This is an illustration of the normal growth and decline of human enterprises. Some companies succeed and others fail, sometimes because of weaknesses in management and sometimes because of overwhelming external pressures. But entwined in the history of Douglas is also the issue of "industrial policy," distorting the workings of unfettered competition. Government investment, which does not require a commercial rate of return; protected markets; and military development contracts that create technology fundamental to commercial airliners all played a role—first in the growth of Douglas and, later, in the growth of Airbus Industrie, which led to Douglas' decline. As we will see, it is difficult to untangle such government intervention from the fate of any of the aircraft and engine companies.

During and after World War II, Douglas designed and built a series of successful attack aircraft for the U.S. Navy, culminating in the AD-1 and the A-4D Skyhawk, the latter a simple and robust aircraft powered by a single Wright J65 and later by a Pratt & Whitney J52 turbojet. (J52 engines provided the core for the Pratt & Whitney JT8D. They are also old and hard to support. The air force of the Republic of Singapore replaced the J52 engines in its Skyhawks with modified GE F404 engines, thereby also improving aircraft payload and range.) But Douglas was much more famous for its airliners, beginning with the 1932 Douglas Commercial No. 1, or DC-1. This aircraft was designed in response to a specification issued by TWA and was quickly refined into the DC-3, at the time the most radical airliner in the world. It had a metal monocoque fuselage structure, an externally unbraced and immensely strong low wing (Douglas demonstrated its strength by driving a steamroller over a wing section without damage), trailing edge flaps, retractable landing gear, and variable pitch propellers, advanced technical features that gave it performance far in advance of any competitor and gave the American airframe industry a commanding lead in the world market for airliners. Thousands were built through the end of World War II.

After the brilliant technical and commercial success of the DC-3, and later of the four-engined DC-6 and DC-7, the 150–250 seat DC-8 was the first Douglas jet transport, a four-engined competitor to the Boeing 707. The DC-8 was a very efficient long-range airliner, and Douglas had a base of loyal airline customers built around the DC-3, DC-6, and DC-7. Nevertheless, Boeing was the first American manufacturer in the market with a jet airliner and was able to build up a substantial lead in market share before the DC-8 became available.

DeHavilland was actually the first in the jet airliner market with the Comet I; Vickers first with the 40-seat turboprop Viscount; Aerospatiale first with the twinjet Caravelle; and DeHavilland again first with the three-engined Trident. But U.S. airlines, collectively the largest single geographic segment of the world market, preferred to buy American aircraft because of their known performance, economy, and reliability, and even more important, the degree of product support offered by American manufacturers located close to the airlines. The American manufacturers designed their airliners with great attention to the specifications of the American airlines, just as DeHavilland and Vickers paid close attention to the requirements of BOAC, a single carrier. The Comet went out of production after a series of disastrous crashes caused by structural problems in the hitherto unexplored regimes of high-altitude flight. None were ever sold to American carriers. A small number of Caravelles and Viscounts were bought by U.S. airlines but were soon replaced by DC-9 and Boeing 727 and 737 airliners.

After the DC-8, Douglas then launched the DC-9, a twin engine 100-passenger airliner for short and medium routes, and this found a very good market in the United States and around the world. But the launch of Douglas' next product, the 250-seat trijet DC-10, sowed the seeds for the subsequent steady decline of the company. Donald Douglas, Jr., the son of the company's founder, was chief executive when Douglas launched the DC-10. He decided to accept no more orders for the DC-8 and ordered the destruction of the production tooling, dies, and jigs for the DC-8 to avoid stealing sales from the new trijet. It was almost certainly a bad decision because there were many airlines that considered the DC-8 the best of the long-range airliners and would have ordered more, without cannibalizing future orders for the DC-10. Boeing, it may be noted, has always continued to encourage airlines to keep ordering its existing airliners, and introduces new models only when the orders for aircraft in current production fall off.

Although the DC-10 held its own against the Lockheed L-1011 Tristar, the price competition between the two drove Lockheed to the brink of bankruptcy and caused Douglas severe financial stress. Donald Douglas Jr. was forced to look for external financing and, as we have seen, finally sold his company to McDonnell Aircraft in 1967 (the combined company was called McDonnell Douglas). With McDonnell's financial backing, Douglas was able to grow the DC-10 into the MD-11 but was unable to find the investment needed for the design of a new wing for a bigger and more capable MD-12. Ominously, neither McDonnell nor potential outside investors such as Taiwan Aerospace were convinced that they could recoup such investment from MD-12 sales, in competition with the Boeing 747, Airbus Industrie A330/A340, and Boeing 777. In the late 1980s, the trijets lost market position to the new long-range twins, the Airbus Industrie A310 and Boeing 767, whose cockpit crew of two instead of three, and two engines instead of three, gave them lower direct operating costs per seat–mile. The twins were well adapted to the traffic patterns of the western Pacific, north-and-south and always within easy reach of an emergency landing site. The aircraft and their engines demonstrated excellent reliability and quickly received FAA approval for passenger flights that would keep them within 180 minutes of an emergency landing site after the hypothetical failure of one of the two engines. Such extended twin engine operations (ETOPS) capability allowed the airlines to use twin engine airliners for trans-Atlantic and ultimately for trans-Pacific routes also—at the expense of orders for the DC-10 (ironically, the DC-10 had grown out of an American Airlines specification for a twin-engine airliner capable of trans-American nonstop operation).

Douglas won a U.S. Air Force competition for the C-17 military transport, the successor to the C-141. Now a division of McDonnell–

Douglas, Douglas encountered long schedule delays and large cost overruns in its development of the C-17. The U.S. Air Force demanded effective corrective action on pain of canceling the entire C-17 contract. In response, McDonnell forced a traumatic reorganization of the entire Douglas team, requiring justification for every position, from president down to the lowest sweeper; and for every person in every position. Naturally, there was a lot of turmoil inside Douglas, and two years had to pass before there was any semblance of stability in the organization. After that, despite the worst fears of the customers, productivity and quality finally began to improve. But the organization had paid a considerable price in the morale of its work force, from the executive office all the way to the shop floor.

In 1991 and 1992, following the war in the Persian Gulf, the European and North American economies went into recession and orders for new airliners dropped sharply. At the same time, in response to the end of the Cold War, the U.S. DOD encouraged the reduction in size of the U.S. airframe industry to match the lower level of military orders, in turn reducing the income available from defense contracts that could be used for the development of other new products. Orders for the MD-11 dropped to a handful a year, and there was no money to develop a competitor for the more modern twins of Boeing and Airbus Industrie. The only arrow left in Douglas' quiver was the venerable DC-9, to which Douglas had already added a new cockpit and new JT8D engines with a higher bypass fan, to woo the replacement market represented by the many existing DC-9 operators. The MD-80, and the subsequent MD-90, were quite successful, and as derivatives of the DC-9 probably cost less to build than the competing 737-300 and A320. They were followed by the MD-95, a version shortened to 100 seats, for which Douglas has not yet received significant orders. The pressure of competition remained relentless, and airlines may have found some advantage in buying from a manufacturer that could offer them a broad product line, something that Douglas could no longer do. At least that was the argument advanced by Airbus Industrie when it solicited funding and approval from its owner companies and governments to launch the 110-seat A319 and a new very large aircraft to compete with the Boeing 747.

The end came quickly. In October 1996, McDonnell–Douglas announced that it was dropping plans for a new Douglas wide-body jetliner of 300–515 passengers; in November, the U.S. Air Force eliminated McDonnell Aircraft from its Joint Strike Fighter (JSF) competition, leaving the field to Lockheed–Martin and Boeing (the JSF is intended as the successor to the F-16 and AV-8B as a light and stealthy fighter for U.S. Air Force, Navy, and Marine Corps and the British Navy; British Aerospace was McDonnell's partner and probably will be involved with either Lockheed–Martin or Boeing in the continuing competition). Later that month, USAir ordered 120 Airbus Industrie aircraft, a mixture of A319, A320, and A321, and announced that it planned to order another 120 later, with options for 160 more. Within days, American Airlines announced its intention to buy 630 Boeing airliners over the next 20 years. Both airlines need to replace older models of the Boeing 727 and 737 and the Douglas DC-9 and MD-80. Both are convinced that there are important economies in standardizing one family of aircraft in their fleet, and both ordered aircraft more advanced in capability and performance than the MD-82 and other Douglas models.

There were reports in the U.S. financial community of merger discussions between Boeing and Douglas, and even informed speculation that Airbus Industrie would forge an alliance with Douglas. In the past, the corporate structure of Airbus Industrie had created many obstacles to a merger with a U.S. publicly held company, but Airbus Industrie announced that it expected to convert itself into an independent corporation with full and public accounts, in place of its current *Groupement d'Interêt Economique,* which allows costs to be hidden in the transfer prices between Airbus Industrie and its four national owners (Aerospatiale in France, Daimler Aerospace in Germany, British Aerospace, and CASA in Spain; there are plans for the Italian firm Alenia to join future Airbus Industrie programs as a partner). Such a reorganization followed by a merger with Douglas could have given Airbus Industrie an American manufacturing presence—an advantage in competing for American sales—and additional production capacity for less investment than expanding its European production. No sooner had Airbus Industrie announced its reorganization ideas than Boeing made public an agreement under which Douglas would join it in the development and production of growth versions of the 747 airliner. Boeing received orders for 618 airliners in 1996, the most in a single year since it launched the 707 in 1955. Ronald Woodard, then still the president of Boeing's commercial airplane group, said, "McDonnell Douglas has excellent design and production capability—both in people and facilities—that are not being fully utilized." Instead of layoffs at Douglas after the decision not to develop a new jumbo jet, Douglas engineers and production workers were to move to Seattle to work on Boeing programs, with some work to be transferred from Seattle to Douglas in Long Beach, California. According to Wolfgang Demisch, a financial analyst with BT Securities, Harry Stonecipher, then the McDonnell Douglas president and chief executive, made it clear that he would not follow his predecessors in selling aircraft at a loss to retain market share. "If that means that McDonnell Douglas has to be a subcontractor to Boeing, he'd rather be a profitable subcontractor," Mr. Demisch said.

Matters did not rest with the 747 subcontracts. On December 15, 1996, Boeing and McDonnell Douglas announced a merger. Boeing had 60% of the airliner market in 1996, Douglas 5%. McDonnell Douglas was second largest American contractor to the U.S. DOD, with orders in fiscal 1995 of $8 billion, Boeing the ninth largest, with orders of $1.8 billion. Obviously their strengths do not compete with each other; indeed, McDonnell Aircraft's long leadership in the development and production of fighter aircraft may strengthen Boeing's bid for the new JSF. Airlines using Douglas airliners will be reassured that they will have continued product support and spare parts from a supplier strengthened by its merger with Boeing, and Boeing will no doubt continue to accept orders for new Douglas airliners as long as there is demand for them at reasonable prices.

Romantics will mourn the passing of one of the pioneer names in the commercial aviation industry. But it was clear by the late 1980s, and perhaps as early as 1970 when Lockheed went into receivership during the competition between the Tristar and the DC-10, that the commercial market could not support all the capacity that existed at the end of the Cold War. The Russians were trying to keep all of their factories open, none of them with enough work to be profitable. Americans were dealing with the problem by merger and consolidation and by eliminating excess capacity. The Boeing–Douglas merger was the most recent step in the process.

Chapter 23

The Cyclical Airline Industry

For years, airline operations were tightly regulated. Following the example of other public utilities, governments encouraged them to recover their costs with ticket sales and set the price of tickets to yield a profit above the costs. There was no effective price competition. Sometimes governments subsidized equipment purchases and the operation of mail and passenger routes deemed essential. Today airlines fly where they find demand for tickets rather than where government policy may dictate. Thirty years ago, the airline industry may have been run by pilots and engineers with a love for aviation technology; today's airlines are generally run by bankers who want only a good computer reservation system and low-cost available seat–miles. They are prepared to lease their aircraft and to subcontract everything else—ticketing, maintenance, luggage handling, and catering. The central question is whether they have learned to cope with the economic cycles of profit and loss that afflict such capital-intensive industries.

The world economy affects demand for airline seats and so do other external factors such as the Gulf War of 1990. In recent years, demand for tickets has been strongly cyclical, with a period of 8–10 years. When demand (measured in revenue passenger–miles) is growing, airlines overorder new equipment. At the peak of demand, when aircraft factories are trying to accelerate their production to maximum capacity, the lead time for aircraft delivery is three years; just as demand drops, the previously ordered airliners arrive. The airlines instantly have large excess capacity and have incurred a huge debt to pay for the aircraft they ordered three years earlier.

Internationally, regulation of competition to shield small airlines has been done through bilateral agreements for landing rights and through traffic pooling, in which two airlines divide the revenue from traffic on a route both serve, no matter which of them actually sold the ticket or flew the passenger. It is not a recipe for efficiency. With the consolidation of the European Union, some of the smaller national airlines have lost such economic protection and are having difficulty attracting traffic, difficulty in reducing their costs, and great difficulty matching lower fares offered by larger and more competitive carriers. Austrian Airlines faces this problem, for example, and so do even larger carriers such as Scandinavian Airlines. Neither Austria nor any of the Scandinavian countries is a national market large enough to justify an airline's existence in purely economic terms, but neither is Holland nor Singapore, where KLM and SIA, by targeting a regional market rather than a national one, have managed over the years to create a comparative advantage for themselves before any of their neighbors could set up in competition. External factors such as the increase in jet fuel cost after creation of the OPEC cartel or currency exchange rate fluctuations also affect economic activity and, therefore, the health of airlines. At one point, for example, Lufthansa faced a severe currency squeeze from incurring its overhead costs in D-marks and earning its ticket revenue set by competition in U.S. dollars.

In the case of national airlines, the capital for new airliners may come from the government, directly for airlines such as Air China, and indirectly when state-owned airlines such as Air France, Iberia, or Olympic receive government funds that wipe out their accumulated debts. However, in most cases, airlines borrow the investment capital from banks, insurance companies, or pension funds in the expectation of paying off the loans out of the profits from ticket sales, a form of financing pioneered by the railroads 100 years ago with so-called equipment trust certificates or chattel mortgages. Some government-owned airlines can receive government guarantees for these loans, enabling them to secure lower interest rates than their own credit rating would justify. Typically they can get such government backing when they sell enough tickets to fill up their aircraft: covering costs is the criterion rather than profits. Other airlines usually need a record of one or two years of profits, no matter how small, before they qualify for loans. When the market is tight and airlines have trouble finding enough commercial money to pay manufacturers, customer sales financing by the manufacturer becomes important. It can take the form of guarantees by the aircraft and engine manufacturers of the airline's bank loans or sometimes of direct financing, the airplanes serving as collateral so that the risk is manageable.

Regional and commuter airlines that operate independently usually lack big financial resources and cannot get bank financing on the same terms as major carriers. Price is critical for them, and manufacturers have to be able to help them find financing for aircraft acquisition. The certifying agencies impose the same safety standards on the small airlines as on the big carriers; of course, the engine manufacturer's quality standards are the same for all.

The economic cycles of the airline business are amplified for the airframe and engine industries, which find themselves at the end of a game of economic "crack the whip." The last 20 years have been a sobering experience for the airlines and the manufacturing industry. The airlines, which used to live in a regulated environment, protected against competition and, with profits almost guaranteed, had become overstaffed, unproductive, and prone to inefficient management of their economic assets. Deregulation in the United States gave them a huge shock. New players came into the field, nimble, with low overheads and cut fares. The established airlines had to learn how to compete or face failure. Most of them managed the transition, restructuring themselves and their routes and forming alliances that allow them to operate worldwide. The international players used to be Pan American Airways, TWA, and Northwest; Eastern, American, Delta, and United Airlines were largely domestic carriers. All of this has changed radically. Pan American Airways and Eastern Airlines have ceased operations, and TWA has shrunk into a much smaller airline. American, Delta, and United Airlines have taken over and expanded their international routes. The survivors have changed their route structure and try to use aircraft of the optimum size and performance for long-, medium-, and short-range routes. The engine manufacturer seeks to serve all of these markets and applications with a few standardized engine models, for example, CFM56 engines for the short-range A320 and the long-range A340, and CF6-80 engines for the short-range A310 and the long-range A330.

The size of the market has shrunk drastically from its peak in 1987. Total combined annual U.S. spending on commercial and military airframes and engines has dropped from $140 billion in 1987 to $80 billion in 1995. U.S. government funding has dropped even more sharply. Advanced technology work continues at its previous level, but development has gone down by a factor of two and procurement by a factor of three. There is no reason to expect U.S. government funding to return to the levels of the 1980s. The simultaneous drastic drop in demand in both the military and commercial engine markets since 1993 is considered unprecedented and unique because the conventional wisdom was that the military and commercial engine markets complement each other—as one falls, the other rises, damping the extreme fluctuations of demand. Data on the U.S. market do not show clear evidence of such a countercyclical balance between military and commercial engines. (The Aerospace Industry Association [AIA] publishes historical sales data compiled from its members and by the U.S. Commerce Department's Bureau of the Census; sales to the U.S. government include sales destined for foreign governments as grant aid or under the foreign military sales process). Between 1950 and 1957, the U.S. military engine market grew by a factor of ten (measured in current-year dollars) during the Korean War and while the U.S. Air Force and U.S. Navy converted from propeller to jet power. During the same period, the commercial market grew steadily by a factor of eight, beginning with the explosive growth of commercial air traffic after the end of World War II and continuing in the mid-1950s with initial orders for the Boeing 707 and Douglas DC-8.

After the jet conversion of the U.S. armed services was more or less complete, military engine sales in 1957–1960 dropped by half, but commercial sales held steady. Between 1961 and 1970, the military market doubled, in support of the war in Vietnam and the introduction of a second generation of fighters and attack aircraft; the commercial market tripled during the same period. During the recession of 1971–1972, the U.S. military market again dropped by half, and the commercial market by only 8%.

From 1973 to 1991, both markets grew enormously, the military by a factor of six, the commercial market by a factor of ten. The commercial market became bigger than the military market around 1973 and has remained bigger ever since. Even after the inflation of the late 1970s is filtered out, this simultaneous growth in both markets is remarkable. There were several factors driving this strong growth: the large military equipment procurement programs started by President Carter and vastly expanded by the two Reagan administrations; the OPEC oil embargo after the 1973 Arab–Israeli war, which raised the price of jet fuel and made new turbofan engines economically attractive; and the new international rules banning the noisy first-generation jet airliners from major airports by a certain date, forcing airlines to place orders for quieter airliners.

Department of Defense procurement began to drop in 1987 to reflect *glasnost* and *perestroika* and then the collapse of the Soviet Union in 1990. In 1991, the Gulf War caused a drop in commercial airline traffic. The war was soon followed by a worldwide economic recession, leading to a further drop in demand for air travel. Airlines had been placing orders for new aircraft on the assumption that ticket demand would continue to grow steadily at an annual rate of 5–8%, depending on how optimistic the forecasters were; when actual traffic fell by a little more than 1% in 1990, orders already in place created an immense excess of capacity. New orders were cut off abruptly. For a mixture of reasons, both the military and commercial markets were contracting at the same time. Despite occasional pump-priming military orders (such as the U.S. Air Force order for 40 KC-10 tankers and a significant number of Learjets when the commercial market for Douglas and Lear had collapsed entirely), there is no clear evidence that the military and commercial markets grow and shrink in a countercyclical manner; in fact, there is some evidence that for most of the past 40 years they have moved in rough synchronism.

In 1992, the world's airlines ordered 700 airliners. The bank analysts who keep an eye on financing for new airliners calculated then that all the world's airlines together would need 400 airplanes a year for replacement of old aircraft and for traffic growth. *Aviation Week & Space Technology*, in its March 23, 1998, issue quotes Boeing chairman Philip Condit estimating an annual order rate of 8% of the world's fleet—5% for long-term growth and 3% for replacement of old airliners. Fewer than 100 jets were scheduled for delivery to the U.S. airlines in 1995 and 1996, the lowest delivery rate in 20 years, but in 1996 and 1997 the rate of new orders for airliners picked up again, to levels not seen since 10 years earlier.

Wars and threats of war depress the demand for air travel. Embargoes and cartels that raise the price of jet fuel steeply raise airline operating cost and depress demand, which is elastic and very sensitive to ticket price. Russia is a special illustration of this: aviation is a widespread mode of transportation, given the immense distances between cities and the relatively underdeveloped road and rail networks. When jet fuel is arbitrarily cheap, Russians fill their airliners. When fuel is priced at world levels and airline tickets are priced to recover the cost, demand drops by an order of magnitude.

Economic recessions, including those caused by high interest rates, depress economic activity and therefore demand for air travel. High interest rates make borrowing money to buy new airliners expensive for airlines, and also make the investment of large sums of money in engine development that will not result in sales until several years later unattractive. The higher the interest rate, the less future income counts against today's research and development expenses. That these are truisms does not make them unimportant. Their impact on the health of an engine or aircraft company can be a matter of life and death. As we have seen, at the end of 1995 Daimler–Benz, the majority owner of the Dutch aircraft company Fokker, announced that it would no longer underwrite Fokker's losses with working capital. Fokker was the victim of a slump in demand for its small airliners. An acquisition made by one Daimler chairman, pursuing his strategy of international alliances, becomes a disposition for the next chairman, concerned with the lack of profits in a stagnant or declining market.

Timing is everything, the cabaret comedian tells us. A few years ago, the Canadian aircraft company Canadair certified its Challenger executive aircraft during a world recession, the exact moment in the economic cycle when orders for such aircraft slowed to a crawl. Canadair went into receivership and had to be reorganized out of existence, its executive aircraft division bought by Bombardier, a successful Canadian manufacturer of snowmobiles and subway cars, for ten cents on the dollar. Bombardier had the capital and the tenacity to wait until demand recovered; today the Challenger is one of the more successful executive aircraft on the market and has been modified by Bombardier into a successful commuter jet airliner. What was a losing proposition for Canadair became a winner for Bombardier.

The market analyst needs to pay attention to such *extrinsic* factors and forecast them as best is possible. Then the *intrinsic* factors of demand need to be considered, illustrated by this not-so-hypothetical example:

Assume that 50% of the U.S. market is business travel, roughly proportional to corporate profits, and that the other 50% is personal travel, roughly proportional to disposable income. If data from the U.S. Federal Reserve Bank and other economic forecasters lead you to believe that corporate profits will grow by an average 5% per year from 1996 to 2000, then the total demand for tickets (measured in revenue passenger–miles) will grow by 2.5% per year from this cause alone. If disposable personal income in the same period grows between minus 1% and 0% per year, demand for personal travel will drop. Because demand is price-elastic, fare reductions could hold it level or even make it grow by, say, 1% per year. But fare reductions decrease financial yields correspondingly, including yields from business travel. The net effect on revenue growth would be zero.

It would be reasonable to conclude that demand for revenue passenger miles in North America will grow by 2.5% per year until 2000. If disposable income grows at the limit of what the U.S. Federal Reserve Bank will allow to control inflation, demand could rise to 3.5% per year. Europe's economic cycle lags that of North America. The "open skies" policy of the European Union will lead to airline consolidations and mergers that will eliminate excess airline capacity so that we can predict zero growth in Europe in 1996 to

Table 1 Worldwide growth in airline travel from 1996 to 2000

Region	Weight in the market, %	Growth rate, %	Net effect, %
North America	40	2.5–3.5	1–1.4
Europe	30	0	0
Asia	30	5–8	1.5–2.4
Total market growth			2.5–3.8%/yr

2000. Asia's growth will continue at a rate higher than in the United States or Europe (we are looking at long-term economic cycles here, rather than the downturn of Asian economies in 1997–1998, which we expect to be followed by continued growth). Table 1 shows how we can add all of the factors. How many new aircraft will the airlines need to satisfy this growth? How many to replace obsolete airliners? How many airliners ordered in earlier years and not used now remain to absorb this combined demand?

The market will be less exuberant than that of 10 and 5 years ago, continuing to grow at a more modest rate, but from a higher base; however, its cyclical nature will not change (that means cycles of layoffs and hiring every 8–10 years). The airline industry, and behind it the airframe and engine manufacturing companies, need to accommodate themselves to that and invest their human and financial resources accordingly.

Chapter 24

Lessons for the Investor

CONSIDER the question faced by Russians anxious to convert their vast and capable aircraft and engine industry from military production to the commercial market; by the French government faced with dramatic losses at SNECMA; by the board of BMW pondering a realistic price to offer for MTU. Indeed, consider the question faced by GE first in 1952, and since then continuously as the market demanded further investment. Is the manufacturing of commercial aircraft engines a prudent investment? What lessons can be drawn from these 45 years of GE's experience?

It is clear, first of all, that you cannot dabble in this business. There must be a basic level of technology for product design and manufacturing processes, and that requires research and development investment at a high level. GE Aircraft Engines spends several hundred million dollars per year on research and development, year after year. Product development is even more capital-intensive. For commercial success you must have your engine on the right airplane, on more than one if possible; sales financing requires substantial capital. The development from scratch of a new commercial engine such as the GE90 requires a multi-billion dollar investment for product design and certification, production machinery and special tooling, the inventory of work in process once production starts, and sales financing to launch the new engine. In addition, the certification process takes four to six years and not until certified engines are delivered to airlines does the manufacturer receive any revenue from sales. Cumulative cash flow for a new engine development program is shown in Fig. 11.

As the manufacturer delivers production engines and airlines pay for them, the negative cumulative cash flow is reduced, until the manufacturer breaks even some 8–14 years after the program is launched. The management consultants McKinsey and Company have estimated that it took Pratt & Whitney 14 years to break even on JT8D development—an engine that had no effective competition in its thrust class. Aircraft engine manufacture needs patient capital! Of course, if competitors offer their new engine models in turn, the manufacturer may have to expend additional development and tooling funds to create and certify better or bigger versions of the new engines and to certify them on other aircraft. That pushes the break-even point further into the future. There is a minimum number of engines that must be sold to break even and a great risk in depending on a single airframe application.

A successful aircraft engine model stays in service for 20–40 years. As a rough rule of thumb, a program resulting in the sale of more than 2000 engines over the first 10–15 years is likely to break even.

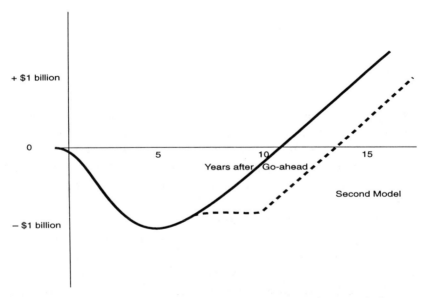

Fig. 11 New engine development program—cumulative cash flow

A big program leads to annual production of 200 engines or more. After the break-even point has been reached, the manufacturer expects to make handsome profits on sales of new engines and of spare parts.

Consider the time value of money. A manufacturer commits an investment of $2 billion, with a peak negative cash flow of $1 billion after five years, in the expectation that five years further on the manufacturer will receive prices big enough in the face of competition to pay for the cost of production, recoup the development investment, and pay interest on the capital tied up. When interest rates are low, that can be an attractive investment option. When interest rates are high, income 5–10 years later is worth less today, and investors may look to other investment opportunities that offer a quicker break-even point.

The engine industry requires large capital and a long period of time to break even—necessary conditions, but not by themselves sufficient for success. It took more than 20 years for GE to establish itself in the first rank of manufacturers of commercial engines, beginning with the less-than-successful CJ805. We learn best from our mistakes. From the unhappy experience of the CJ805, GE learned some of the other conditions for a successful commercial engine business:

- A superior *product*, closely matched to the airplane.
- Superior *product support* and the correcting of problems effectively.
- The *staying power* to support the product and the customer through the ups and downs of the business cycle and the world economy.

Management, according to the old management joke, sets objectives and assembles the resources required to attain them and then gets out of the way of the workers. The first step in such a long-cycle business is product planning, a process that matches engine concepts to the likely needs of the airliner market 10 and more years into the future; it is product planning that defines superior engine designs. Being first in the market with a good product is an inestimable advantage, hard for the competition to overcome. Aggressive product planning defines aircraft engines with a clear technical advantage over competitors in terms of thrust, installed fuel consumption, life, reliability, and maintenance cost, all visibly of direct benefit to the airline's profitability. While such a performance advantage reduces the pressure for a lower price or other concessions, manufacturing cost superiority is equally important, generating the opportunity for profit on engines and spare parts even in the face of fierce price competition. Low cost begins with the mechanical design of the engine, and requires aggressive investment in advanced manufacturing technology as well as in design technology (Kawasaki won a manufacturing order from Boeing for the 777 fuselage over Lockheed and Northrop by investing in the development of a huge horizontal milling machine with computer-controlled holding fixtures—greatly increasing the speed and flexibility of production and eliminating hundreds of separate holding fixtures for different fuselage sections). The manufacturer can never rest on previous successes but must constantly develop new product and manufacturing technology to maintain leadership, not technology for its own sake but technology focused on a product 10 years in the future. Starting with the product plan, there must be a road map of technology needed to allow the product plan to be realized.

Airlines compete against each other every day and cannot survive long with an inferior product. At the same time, the airline industry is technically conservative and averse to risk. It is skeptical about claims of new "advanced" technology and wants to be comfortably sure that this will work as advertised. Airlines generate data on competitive engines every day they fly—a comparison they trust more than the brochures of the marketing teams. To break into the market against the entrenched JT9D, for example, GE had to offer the airlines better engine parts life, lower maintenance costs, and lower fuel consumption, and, until all of that could be proved in service, to back up the offers with financial guarantees (on a 747, a 1% advantage in specific fuel consumption can reduce fuel costs by $150,000/yr; actual fuel consumption in service is affected by aircraft drag as well as by engine efficiency so that identifying the precise cause of a shortfall is not an easy matter). In this industry, there is little room for complacency, and none for underestimating competitors. Capable competitors learn from each other quickly, in technical matters and in commercial terms. GE's engines are as good as their competitors', often better, but seldom by much. GE's competitors also make good products and, during a competition, any one of them can make financial concessions to make performance differences fade away. In the 1960s, noting that its competitors were vulnerable in the area of product support, GE saw a way to gain prominence by developing a reputation for excellent product support and always delivering what it had promised. The airlines, of course, were happy to have more competition among the engine companies, and better support.

GE also saw an opportunity to gain a market lead by entering into global alliances, calling the strategy "Share to Gain." Rather than insisting on making every part and every technical or marketing decision itself, GE was prepared to share management, engineering, and access to production with partners who could bring their own advantages to a program: preferential access to markets, development funding, and unique technology. Concluding that a less-than-100% share in a winning global partnership could be bigger than 100% of a smaller GE-only program, and certainly bigger than an opportunity lost to a competitor, GE picked its partners with care: in France for government sponsorship of the engine industry; in Japan and in Germany for their industrial power; and in smaller alliances in Italy, Sweden, and Britain that addressed those markets and also served to limit the options available to Rolls–Royce.

This could not be a short-term proposition. It was a commitment for the life of the product, 30–50 years, and the result of intense discussion between the top levels of GE and the airframe companies and airlines, who had over the years all developed close personal relationships. An airline stakes its future on any new equipment it buys, and it must be confident that the supplier will be there to support the product for the full length of time it is in service, not only with manuals and spare parts, but even more importantly with technical support when the inevitable problems arise, including, in the worst case, a willingness to share the financial burden. A major engine problem, such as a titanium fan disk failure, confronts the airline with large and unforeseen expenses. It expects the engine manufacturer to share the burden financially as well as to correct the problem technically and logistically. This goes beyond the concept of warranty, the manufacturer's maximum contractual liability: serious problems can cause the airline expenses far in excess of the warranty. Airlines expect the manufacturer to participate in such expense, not necessarily with full reimbursement but at least to share in it. There is a tacit understanding in the industry that there is a sort of partnership between the airline and its supplier, in which they share the consequences as well as the benefits. When GE entered the business, the airlines asked for—and received—such a commitment from GE's corporate management. The managers who run these companies stamp their personality on the business. Establishing close personal relationships with them at all levels of the management structure is very important. GE worked hard to build credibility with its customers at all levels, from the chairman to the line mechanic.

Above all, the manufacturer must be able to produce what it promises. If a manufacturer fails to deliver on time, its product does not meet performance specifications, or if there are serious quality problems or the engine does not meet its maintenance cost guarantees, the financial penalties can be very severe. Airlines will not accept such problems and will look for compensation. Future sales are of course also at risk. Whenever there was a problem, GE rushed out to tell the customers as quickly as it told them about its successes.

All of this directly affects the mechanical design of the engine. High-value engine parts must be designed to meet the airline's life expectations and must be capable of being repaired several times. Repaired parts must function like new parts, not as durably, perhaps, but equivalent from the point of view of performance. When it began, GE had a lot to learn about repair procedures, restoring hardware to recover performance, and the logistics of managing hardware after the initial engine delivery.

Changing engines or even parts at short intervals as a way of attaining high reliability is not acceptable today. The FAA limits on how long a twin-engine airliner may fly over water after an engine failure before it reaches an airfield determine how far away from land the aircraft is allowed to fly and, therefore, its route between continents. The higher the demonstrated engine reliability, the longer the allowable limit, and therefore the straighter the course over the ocean. The rules for ETOPS mandate an in-flight shut-down rate of less than 0.02 events per 10,000 engine-hours; with engines of such high reliability, twin-engine airliners routinely fly across the Atlantic and Pacific oceans on routes for which trijets and four-engine airliners were formerly used, offering the airlines big cost savings. During the past 25 years, as engine reliability has grown, airlines' expectations have also grown more than anyone could have imagined. To simplify their operation, training and maintenance, and the logistics of spare parts, airlines want the same engine model for twin-engine ETOPS and for four-engine aircraft. They may be prepared to bring an engine into the overhaul shop to replace a combustor or some turbine components, but they do not want to overhaul a compressor for 10,000 hours or more—three to four years of operation. Amazingly, the engine industry has proved itself capable of attaining such levels of equipment life and reliability.

A big fleet of engines in service needs a large supply of spare parts, even with the growing durability and reliability of parts. Engine manufacturers like to sell spare parts to the airlines so that the airlines are ready to meet their own overhaul needs quickly out of their own inventory. The larger the fleet to be supported, the easier it is to make good statistical forecasts of the spare parts required. But holding an inventory is expensive, and airlines prefer that the manufacturers carry the inventory. Recently, to dress up their balance sheets, some airlines and perhaps manufacturers also have turned to third parties to invest in spare parts inventories. There is a price paid for that because the cost of carrying inventory does not disappear. The form that this cost then takes in an airline's books is as a higher price for each part, but not as a large sum of working capital tied up in inventory.

Revenue from spare parts sales is a big factor in the profitability of a product line and in accumulating the resources for the next generation of engines, which will be financed from the successful sale of the previous engine and its spare parts. The U.S. government has contributed some funding to research on commercial engine technology, small in proportion to product development costs. Among these have been two NASA-funded projects, "Quick See" and "E Cubed," described briefly earlier. Later, NASA contributed some funding to advanced prop fan development, and to component work for the next generation of supersonic transport engines, the High Speed Civil Transport project. In general, the U.S. government asks for a royalty on commercial sales arising out of the use of technology it has funded.

Commercial and military engine programs both put great emphasis on low cost, but the manner in which they do so is quite different. The U.S. military services negotiate what they consider tough cost targets annually, with stringent cost reductions from year to year, the manufacturer being measured by how well he achieves those targets. In the commercial program, the manufacturer begins with a selling price that is usually far below the cost of the initial engines, requiring the manufacturer to build in cost reduction plans at the very beginning of the program so that it can ultimately sell engines at a profit and recover development expenses. In a military program, the customer tells the contractor what to do. The contractor does not make many decisions alone, but instead submits ideas to the customer. If the customer wants to accept a design change, it does so; if it wants to buy spare parts, it does so; if it wants a new gadget on the engine, it issues a contract amendment for its development and manufacture. The designer can propose a change that makes the engine safer or better, and the customer may decide not to incorporate it for its own reasons. In contrast, for commercial engines, any change and all problem-solving are the manufacturer's responsibility. In a military program, the manufacturer spends the customer's money; in a commercial program he spends his own.

The risk and mindset for the engine program manager are thus quite different between the two markets. When public money is being spent, every decision must be traceable, which is why the military services have many more people managing their programs than do commercial customers. For every one of the military managers, the engine manufacturer must have a counterpart. The military services ask for special contractual and technical documentation, and the contractor must comply in detail with the customer's specifications; must demonstrate that compliance to the customer's satisfaction; and furnish all the documents, reports, and data stipulated in the contract, including data on costs. Should failures or accidents happen after that, the contractor can claim that it has done what was required and that the customer has formally agreed that this was so. Remedial action, if it is required, becomes the subject of a new contract.

This is all extremely expensive and inappropriate for a commercial engine. All an airline wants is a reliable engine at a low price, an engine that will run for a long time. The risk is all on the manufacturer, whom no amount of contract language will protect if 300 people die in an accident caused by the engine. Decisions made by the commercial engine program manager can have catastrophic financial consequences. In the early days, GE's commercial engine management team was young and brave, but it had to give serious and explicit consideration to the risks and potential liability to the company inherent in its decisions. As a small example, GE developed a smokeless combustor for the CF6. No such requirement had been specified by the U.S. Air Force or U.S. Navy, but the flying public's attention to the environment had made it a political necessity for a commercial engine. Every development program faces the prospect that design engineers do not achieve all that they expect: in the case of the old reliable Pratt & Whitney JT8D, failure of an overhauled combustor led to a serious incident with a 737 airliner. What risk did the new smokeless CF6 combustor pose? (Readers can be reassured: The CF6 smokeless combustor has performed safely as long as it has been in service; the U.S. Air Force did not adopt the smokeless combustor in its TF39 engines until much later.) The designer must always consider that a new design could fail catastrophically, damaging or even bringing down the aircraft, or that ejection of shrapnel fragments from a failed part through the casing of the engine could injure people in the aircraft or on the ground. If that happens in a military aircraft, government ownership provides some shield to the manufacturer against liability claims; no such legal shield exists when the engine is used in a commercial airliner.

Luck plays a role, along with skillful management. Some strategic decisions lead to big rewards or big penalties. One can identify a number of such moments in the history of GE Aircraft Engines:

- The decision to use single rotor compressors and aft fans. Pratt & Whitney made a different technical decision: its dual rotor

and front fan allowed Pratt & Whitney to overtake GE as the market leader in the 1950s and hold that lead for almost 30 years.
- The later GE decision to build a dual rotor front fan of very high bypass ratio, which put GE back into the large commercial engine business in the mid-1960s.
- The decision for revenue-sharing with SNECMA and MTU, which helped make GE the launch engine for Airbus Industrie.
- The decision not to launch a successor to the CJ610 and CF700 in the mid-1960s, which allowed Garrett to take over and dominate the business jet engine market.
- The decision to go into partnership with SNECMA to develop the CFM56 as a replacement for the Pratt & Whitney JT8D.

Clearly, some of these decisions were successful and others not. Like GE, Pratt & Whitney also made several strategic mistakes: it did not believe that GE could displace its military engines in the F-14 and F-16 fighters or that GE could make the CFM56 joint venture with a French company succeed, and it believed that the JT8D monopoly would go on forever. There is a risk in believing only what one would like to happen.

The American aerospace manufacturing industry experienced a meltdown in demand after the end of the Cold War with the former Soviet Union. Military orders received began to drop in 1987, over the next 10 years by as much as half from their high point in 1986. Commercial orders dropped worldwide immediately after the Gulf War in 1990 and did not revive significantly until 1996 and 1997. The American industry accommodated itself to these facts by changing its structure, shape, and size. The number of prime manufacturers was reduced drastically by merger:

- Lockheed with General Dynamics.
- Martin–Marietta with GE Aerospace.
- Lockheed–General Dynamics with Martin–Marietta.
- Northrop with Grumman.
- Raytheon with E-Systems and Hughes.
- Rockwell and McDonnell–Douglas with Boeing.

- Lockheed–Martin with Northrop–Grumman (the U.S. Department of Justice has not approved this merger, which will not be allowed to proceed unless the two companies divest themselves of specific now-competing divisions).

Most important, overall capacity was reduced until it realistically matched demand. Boeing, for example, has matched the down-and-up cycle of demand for airliners in its traditional way—by laying off people in the early 1990s when demand dropped, and hiring them again in 1996 as demand rose. Most of the Douglas airliner product range simply ceased to exist after the merger with Boeing; only unique Douglas products will be continued. The DOD encouraged such consolidations and agreed to accept some of the costs of mergers and employment reductions in its contracts; the Department of Justice gave the consolidations benign antitrust approval because competition between two firms remained in most categories and foreign competition remained vigorous. Many of the smaller subcontractors went into entirely different work or simply went out of business. That matches the desires of the purchasing departments of the prime contractors, which have been driving the vendor infrastructure relentlessly to lower its costs. One widely used strategy was to reduce the number of vendors, often from several thousand to several hundred, and to offer the survivors long-term sole-source contracts with long-term price reductions negotiated from the beginning.

Life has not been an unbroken string of successes for any of the three big engine companies. All of them have had their moments of far-sighted technical planning, occasional technical blind alleys, and commercial failures as well as successes. Their competition has been good for the airlines, the airframe companies, and the engine industry itself. All of them have generated substantial profits over the long haul. GE Aircraft Engines has been profitable in *every year except one,* when there was a nationwide strike against the company, and consistently able to finance its own development work out of the profits of its own sales. Because of the industry's enormous entry cost and distant break-even points, there are few successful commercial engine manufacturers in the world. Once in, what are the prerequisites for survival? A clear vision of the technical requirements of the future, the human and financial resources to survive until cumulative revenue exceeds cumulative expenditures, the psychological tenacity to hold on during the downturn of the economic cycle, and sober realism during the upturn.

Chapter 25

Are There Any Lessons for the Russians?

THE Soviet industry has faced an even more drastic reduction of demand, aggravated by competition in its commercial sector from more productive American and European aircraft. Russian aircraft designers blame their problems on the state of the art of Russian engines, but in fact a general liquidity crisis in Russia and the other former Soviet republics has made it difficult to match even internal demand for commercial aircraft with internal production. Some of the Russian factories have tried to diversify into nonaircraft products, such as outboard engines and garden tilling machines, but the prime effort of the major factories has been on exporting what they make now—the military aircraft they have in production—to keep their factories alive. China, Malaysia, and Peru have bought MiG 29 or Sukhoi 27 fighters and so has India, which plans to produce them itself. The export numbers are small and their competition remains the American F-16 (sometimes also the F-18 and, more recently, the Rafale, the SAAB Gripen, and the Eurofighter). During the Gulf War, U.S. combat aircraft demonstrated decisive superiority over Soviet-designed aircraft, although that may have had as much to do with the capability of air crews and command-and-control systems as with raw aircraft capability. Russian exports of commercial aircraft have been minimal, hampered by their perceived low productivity and the genuine lack of product support.

This book had its inception in the Russian question at the 1990 Moscow propulsion conference, "Where does the money come from for new programs?" Looking for an answer has taken us not only through GE's history but also that of the industry as a whole. What lessons can the Russians draw from all of this? That the conversion process takes a long time? They have been at it for just as long. That the market demands good products? They consider their engines as good as any. What is it, then, that distinguishes this history of GE's activities from that of any Russian producer of military or commercial engines? There are two main threads to GE's success. Recounting them makes them sound banal, but they are important nevertheless: the first is relentless attention to cost; the second, that it is vital to form a genuine team with the customer.

The Russian aircraft engine industry grew up in a system where basic research was conducted at research institutes such as the Central Institute for Aerodynamic & Hydrodynamic Research, the famous TsAGI, which corresponds to the aeronautical side of NASA but also performs some of the advanced product planning functions done in the United States by the engine companies themselves, and on occasion builds and tests demonstrator engines to validate a concept. Whenever the time came to define a new aircraft engine, several design bureaus, all of which have small factories for making prototypes for development testing and type certification, were invited to define competing designs based on TsAGI's new technology. When a design had been selected and type certified, the Ministry of Machinery made drawings and specifications available to one or more of its big production factories, which are sometimes closely allied to a particular design bureau and sometimes not. The factories are responsible for production engineering and tooling and, while they can make design changes to improve the engine or to make it easier to produce, the responsibility for approving major changes remains with the design bureau. There has been a certain distance between the responsibility for the initial design and the responsibility for manufacturing quality and cost. Finally, it was the Ministry of Machinery that ordered production engines and provided funds for working capital and the human resources required by the big production enterprises. The design bureaus did of course interact with the airframe designers and technical experts in the air force or in Aeroflot to develop new engines capable of meeting military performance specifications, but there was very little direct contact between either the design bureaus or factories with the actual users of the engines—the operating squadrons or airlines—and very little contact with the mechanics responsible for maintenance. The interaction that counted was with the Ministry of Machinery.

In the past, designers of Russian and American engines have put radically different emphases on maintenance. The Russian philosophy of logistic support was to ensure engine reliability by changing engines at fairly short intervals: as low as 150 hours for military engines and as low as 500 hours for commercial engines. New or freshly overhauled engines are then installed in the aircraft, and the engines taken off the wing are sent to the factory for a complete overhaul. At least one reason for this design philosophy was the wide variation in physical properties of the materials available to Russian designers. Their average strengths were in no way inferior to those of the alloys used in the West, which are, however, manufactured to much tighter tolerances of composition and impurities to permit safe design closer to the average strength. Because of the wide variation of properties around the mean, Russian designers chose to apply large safety factors and to design for maximum allowable loads significantly lower than the average strength of their materials. They also applied such safety factors to the allowable operating life of engines and their parts.

Russian engines were designed to permit quick removal from the aircraft but not for maintenance in the field. The flanges on their casings, for example, were light and mechanically efficient but not sturdy enough to allow easy disassembly at an air base because such an operation was not part of the logistics support plan. Disassembly was done at the factory, which had special tooling for the job. There, during overhaul, the engines are taken apart and every piece is cleaned and inspected. Parts showing any type of damage are replaced with new parts or reconditioned parts, some of which must be ordered, while others may be available from production of new engines. Finally the engine is reassembled, tested, and returned to the operator. Such a process takes a long time, perhaps more than one year from the removal of an engine from an aircraft until it is returned to the operator. The process also requires a large inventory of spare engines: for each installed in an aircraft, there is probably one in the factory for overhaul and one or two more in the pipeline between the operator and the factory. At every major Russian airport, a large number of airliners could be seen parked at the edge of the field, available as spares. That is the most plausible explanation for the massive engine production and overhaul activity noted by Western observers in the Russian factories during the 1980s, far larger than could be explained by conventional production of aircraft alone. This large production capacity became an equally large overcapacity the moment demand for new military engines fell drastically after *perestroika* and the end of the Cold War.

The large number of spare engines of the old system is a perfectly sensible way of ensuring reliable engine operation—as long as inventory is "free" and the operator faces no interest charges on the working capital tied up in spare engines. The Soviet factory's "customer" was the Ministry of Machinery, and the factory responded to the Ministry's schedules and budgets for production; somewhere along the way, the air force or airline received equipment from the Ministry of Machinery. GOSPLAN, the state central planning agency, was responsible for making sure that ministries ordered and factories produced what "customers" of one kind or another needed. There was no close link between the actual user of aircraft engines and the maker of the engines, and there was no capital cost for inventory charged to anyone in the supply chain.

In the United States, airlines must pay for spare engines and inventory is not free. Companies must pay interest for working capital, which includes the price of a pool of spare engines and spare parts. Spare engines require warehouse space and insurance and may become obsolete before they are used. All of these are arguments for maintaining minimum inventory and why mass production companies in the United States and Europe have adopted the Japanese invention of "just in time" inventory. Reducing their inventory of work in process allows them to reduce costs by eliminating the carrying costs of inventory. One way to do that is to keep engines on the wing of an aircraft for as long as possible, a strong incentive for designing engines for long intervals between overhaul. Engines can be equipped with diagnostic systems that signal impending failure or excessive wear of a component, which can then be removed and sent to an overhaul depot or the manufacturer for repair. With the component replaced, the engine as a whole may remain installed in the aircraft, which may remain in revenue service. It is not unusual for an engine to remain installed for 10,000 hours of operation—three to four years of intensive airline service. Some CFM56 engines have remained on the wing for more than 15,000 hours. Such a maintenance philosophy requires that components be designed for long intervals between inspection (that is equally true of Russian design practice), systems be designed to give gradual warning of wear or degraded performance, and parts and major assemblies be designed to facilitate easy removal and replacement in the field with the engine installed.

It is the customer's cost of operations that drives such a design philosophy. In a market economy, airlines will select engines that give them the lowest life cycle cost—initial purchase price plus the cost of fuel and maintenance plus the cost of replacement parts and parts repair. In a planned economy, airlines can only take whatever is the best product selected by the planners. Often that means only the lowest first cost, without explicit consideration of maintenance, spare parts, or even fuel consumption. Airlines that pay a price for jet fuel lower than the world market price have less need for fuel-efficient engines than airlines that operate in the open market.

The engine manufacturer's profit depends on costs lower than its selling price. Low costs are the result of simple mechanical design that keeps down the number of parts, that keeps manufacturing methods in mind; the result of advanced and efficient manufacturing methods; and the result of a productive work force. All of these in turn depend on the manufacturer's knowledge of real market-level costs, including a cost for the capital employed in the manufacturer's business.

No responsible aircraft engine manufacturer, in Russia or any other country, sets out to be deliberately inefficient. But you measure what is important to you, and results follow incentives. For more than 70 years in Russia, production was important and profits were not part of the equation. Low costs depend on the manufacturer's ability to measure real costs unaccounted for under the socialist system, and incentives such as profit to lower them. For the Russians, this kind of economics is a new discipline.

Russian engine designers of course work closely with airframe designers and operators of the aircraft, but that is not quite the same as making oneself a member of the customer's team. As we have seen, this is a vital ingredient in the success of Western aircraft engine manufacturers, who, once they have identified an airline operator as a customer, orient their entire management process to that customer. The engine manufacturer begins by establishing close personal relationships so that designers have a real understanding of the technical and economic needs of the customer and the tradeoffs the customer may be prepared to make in one area for gains in another. The customer in turn gets to know the manufacturer and understand that one can rely on the data and the commitments made. Engine design becomes an iterative process, with periodic review by the customer for diagnostics, maintenance and repair capability, and validation of performance calculations by actual engine test. By the time a new engine receives its type certificate, the airline's engineers are thoroughly familiar with it. The engine manufacturer trains the customer's engineers and technicians and writes technical manuals subject to their editing, while its own field service engineers become members of the customer's technical team. During initial operations, they may meet every flight and review engine performance and any malfunctions with the flight crew, reducing the frequency of such meetings as both sides gain experience and confidence. In the airline's engine shop, the engine field service engineers collect data from the airline concerning engine problems. They monitor their frequency and severity and report them daily to the project manager in the engine factory. They track inventories of engines and spare parts and provide liaison between the airline and the factory on repair procedures—particularly those that the airline itself can carry out—and on any other technical and logistical problems. Modern communications help here: field service engineers at various airfields can be linked to the engine factory by e-mail; spare parts inventories and entire technical manuals can be revised every three months with CD-ROM computer discs instead of file cabinets full of paper manuals. It is a lot more effective to equip the field service engineer and line maintenance mechanic with laptop computers than to try to keep their technical manuals up to date with revised paper pages.

Periodically, the engine product support organization holds a conference at the engine factory for engineering representatives of the customer airlines. At such conferences, the engine company presents a list of *all* problems encountered by *all* customers, classified by the frequency of occurrence and severity of consequence. Engineering representatives of the engine company (in the Russian system, representatives of the engine design bureau would also be present) discuss each problem, proposed corrective action, its status, and logistic status of the active fleet of engines. The engine company's managers invite each airline's representatives to give them a frank presentation of their concerns and try to develop a joint plan of action during the conference, if possible, or within a few days of its end. The objective here is not to generate a warm feeling between the manufacturer and his customer but to influence the engine's design and manufacture by constant interaction and feedback between the engine builder and the user.

There has been significant movement in Russia in the direction of the market. The Russian space complexes such as Energiya have set up joint ventures with U.S. companies to adapt and produce Russian rocket motors for NASA; Boeing has set up an office in Moscow to get aerodynamic design and analysis done by TsAGI and its scientists; the French companies Aeospatiale and SNECMA are working with the MiG design bureau on a new jet trainer to succeed the Franco–German Alphajet. One unique export success has been the carrying of outsize cargo in Antonov An 124 transports; Air Foyle, a British freight forwarder, is the chartering agent. Such joint ventures with European, American, or East Asian firms may turn out to be the most promising strategy for the Russians, coupling Russia's superb test facilities and undoubted expertise in aircraft and engine design with the responsiveness to the market of the external partner (for example, between Russian helicopter enterprises such as Mil and Kamov and American and European companies such as Bell and Agusta). At the moment, Russian design and production organizations enjoy a temporary cost advantage because their real wage rates are significantly below those of the rest of the industrial world and their overhead costs are the subject of guesswork and arbitrary price setting. But that advantage will fade as quickly as it has in China.

The Russian industry knows very well how to design and build high-performance aircraft engines. Its military aircraft are proof of that. What it needs to learn is what GE learned over a number of years: to pay strict attention to cost and its real customers, the airlines.

Chapter 26

The Bottom Line

IT IS easy to become caught up in the excitement of the aviation industry—product planning, the development of new aircraft and engines, the cut-and-thrust of airline competitions. The questions remain: Why did GE enter the aircraft engine business in the first place? Was it a wise decision? Is it still a good business for GE?

It was the U.S. government, in the person of General "Hap" Arnold of the U.S. Army Air Corps, who, because of GE's unique technology in aircraft engine turbosuperchargers, asked GE to undertake the development and production of the Whittle aircraft gas turbine in 1941. Thereafter, GE developed the J33, J35, and J47 engines with U.S. government funding, and went into large-scale production of the J47 and later J73, J79, and J85. These programs were very profitable for GE because of their volume, even though the net profit per dollar of sales was quite low. The U.S. government provided working capital through progress payments so that the return on GE's investment was quite high throughout the war years. The situation was the same for Pratt & Whitney and the other American engine companies.

GE saw the opportunity to become a leader in the commercial engine business, a field that fit its overall business strategy: huge investment is needed for basic research, product development, and production tooling. That limits the number of competitors to a few companies with deep pockets of their own, or with a benign outside investor willing to underwrite them, with the patience to wait several years to break even. GE had unique gas turbine technology, probably better than that of its competitors. It had the technical people and financial resources to create the product and was prepared to wait until its new engines could establish themselves in the market place. What GE lacked and had to acquire was a thorough understanding of what the airlines expected from a supplier and a reputation with the airline customers.

We have seen what counts for success in such a capital-intensive and long-cycle business: continuous investment in research and development; sensitivity to the market and responsiveness to the customer; a superior product of the highest quality; competitive price, inevitably below the cost of the first several hundred units produced; and finally, a deserved reputation for first-rate product support. The lessons are as applicable to the rest of the world as they are to the Russians.

Now, more than 50 years later, GE is one of the leaders of the industry, perhaps the biggest of the engine companies. It has established this position with funds generated from GE Aircraft Engines sales, investments by its partners, and technology flowing from programs funded by the U.S. government. GE Aircraft Engines has been profitable for all but one of the years of its existence and, in sales and profits, it is one of the largest components of the company. Its profits have been remitted to the parent company and also reinvested in new product development. All of the manufacturers like to boast about the benefits their engines bring the customer. The benefits have also flowed to GE's partners and the U.S. government. Their employees too have benefited—GE's chairman Welch once joked that GE Aircraft Engines was an employment agency for engineers and technicians. True or not, they made good products for a fair price and a profit for the company.

That is a decent outcome.

Index

A
Abington, Richard, 210
Adamson, Arthur P., 163–165, 168–171
Airbus Industrie, 53–59, 65
Airline industry economics, 309–316, 317–328
Alfa Romeo, 100–106

B
BMW, 271–277, 286
Boeing
    717 development, 270–277
    Chinese market development, 188–238
    design of the 707, 15
    engines for 747, 61–63
    international market, 65–75
    merger with Douglas Aircraft, 301–308
    Russian market, 250
    turbosupercharger development, 5–6
Brown-Boveri, turbosupercharger development, 3–4
Burgess, Neil, 55–58

C
Chang, Frank, 226
Chang, Walter, 180–238
Cheng, Xiao Shao, 223
China
    GE Office in Beijing, 206–221
    market development, 173–238
    parts production, 194–200
Civil Air Administration of China (CAAC), 173
CJ805 program, 16–22
Curtiss-Wright. See Wright Aeronautical

D
Dassault, 79–80
Douglas Aircraft, 42–51, 53–63
    DC-9 development, 170–171
    international market, 65–75
    MD-95 development, 270–277
    merger with Boeing, 301–308

E
energy-efficient engine, 165–167

F
FIAT, 100–106, 156, 288–290
Foreign Corrupt Practices Act, 119–124

G
Garrett Turbine Engine Company, 29–30
GE Aircraft Engines. See General Electric Company
General Dynamics, 85
General Electric Company
    "Quick See" engine development, 163–165
    Beijing office, 206–221
    business jet product support program, 27–29, 50
    CF6 program, 49–51
    CFM56 program, 132–141
    Chinese market development, 173–238
    CJ805 engine program, 19–22
    DC-9 development, 170–171
    early engine development, 1–22
    energy-efficient engine, 165–167

340 INDEX

Foreign Corrupt Practices Act rules, 121–124
Italian market development, 102–106
J73 engine, 11
Japanese market development, 143–159
LM2500 development, 288–292
military market share, 77–96
product diversification, 279–292
Russian market development, 239–256
service, 293–300
smaller airliner market share, 125–132
TF39 development, 36–37
unducted fan development, 167–171
Gormley, Robert, 224
Grumman Aircraft, 280

H

Hellenic Business Development & Investment Corporation, 115–117
Hemsworth, M.C., 166–167
Huamin, Cheng, 200–202

I

International Aero-Engines, Ltd. (IAE), 156–159
Ishikawajima-Harima Heavy Industries, Ltd. (IHI), 150–159
Italian engine industry, 99–106

J

Japan
  aircraft market, 143–159
  commercial engine experiments, 151–153
  military aircraft production, 149–151
Jets
  business, 23–30
  engine development, 10–22

K

Klockner-Humboldt-Deutz, 285–286

L

Lear, Bill, Jr., 24

License agreements and offset programs, 107–117
Liming Machinery, 191–238
Linton, Robert, 217, 231
Lockheed, L-1011 wide-body trijet, 42–51

M

Maintenance and service, 293–300
McDonnell Aircraft establishment, 23
McDonnell, James S., 23–24
Military market share competition, 97–106
Mitchell, Mae, 210
Moss, Sanford E.
  gas turbine engine development, 2–5
  turbosupercharger development, 2–5

N

Nath, Rajendra, 222
NATO, relationship with engine production, 80–82
Neumann, Gerhard, 13–14, 24–27, 33, 40–51, 54–58, 129–132, 284
Northrop military market share, 77–96

O

Offset programs and license agreements, 107–117
Olson, Robert, 180, 219

P

Parker, Jack, 15, 288
Paterson, Pat, 178
Piaggio, 102–106
Pratt & Whitney, 1, 6
  Chinese market development, 187–238
  commercial engine development, 31–37, 39–51
  energy-efficient engine, 165–167
  Japanese market development, 154–159
  jet engine development, 12–14
  military market share, 77–96
  Russian market, 250
  smaller airliner market share, 126–132
Product diversification, 279–292

Product planning and partnership, 257–267

Q
Quiet clean short-haul experimental engine ("Quick See"), 162–165

R
Research and development costs, 161–162
Revenue sharing concept, 54–58
Rolls-Royce
　Chinese market development, 186
　early engine development, 9–22
　financial receivership, 46–47
　Italian market development, 104–106
　Japanese market development, 154–159
　Russian market development, 239
Rowe, Brian, 212–217, 224
Russian market development, 239–256, 329–336

S
Service and maintenance, 293–300
Smaller airliner market share, 125–132
SNECMA
　CFM56 program, 132–141
　Chinese market development, 222–223
　Russian market development, 241–256
　smaller airliner market share, 128–132

Sun, Paul, 221
Supersonic transport development, 32–37

T
Trippe, Juan, 23
Turbojet engine development, 8–22
Turbosupercharger development, 2–7
Turnbull, Robert, 225

U
unducted fan (UDF), 167–171

V
V-12 Liberty engine, 1–5
Van Duyne, Walter, 50–51
Voeller, David, 237

W
Welch, Jack, 210
Welch, John F., Jr., 253
Whittle, Frank
　invention of gas turbine engine, 7
Wilson, Thornton ("T"), 62–63
Woll, Edward, 84–85, 182
World War II, impact on engine development, 53
Wright Aeronautical, early engine development, 1–22
Wu, John, 210